APHELION

by

Dr. Jeffery Austin Clements

First Printing, January 2015

Library of Congress Control Number: 2014922236
CreateSpace Independent Publishing Platform, North Charleston,
SC

ISBN-13: 978-1502789310
ISBN-10: 1502789310

ACKNOWLEDGMENTS

One never knows how things are going to come together. A simple dream emerging from sleep that captured years of experience, exposure, and insights in science and technology has evolved into this fascinating story. I am indebted to many who impacted me with the substance of this story. Early discussions with Mary and James Chitty about my lifelong interest in science and my technical career helped initiate the story concept from the powerful dream I shared with them. Inquisitive and thought provoking discussion with Norm Langley and his wife Dr. tr Porter and Emelita and Gary Johnson provided further motivation to document the mysterious science and technology behind the story. Early fascination from friends Roz Stevenson, Carol Hall Holliday, LaRhonda Amos, Dr. Joe Oliver, Marty Woods, Erika Puzik and James and Sandra Penn solidified the decision to go forth with the story.

Comments from my wife's sister Ambassador Ruth A Davis nurtured the motivation to document this tantalizing story. My brother Walter's oldest daughter Laura Clements looked at the early draft and encouraged me to refine and publish the story.

I want to thank the initial editors of content and substance Karl Monger, William Greenleaf and Diana Schramer. They provided some hard hitting comments about clarity, character development and several other aspects.

I benefited greatly from discussions, comments and motivation from my scientific friends including Albert Council, Ken Brown, Chris Worthams, D. Kenneth Richardson, James Robinson and Delvin Walker. The religious views captured some of the sermon material from Dr. Michael Beckwith and Reverend Leo R.

Thomas. I should also thank the leaders of the American Institute of Aeronautics and Astronautics (AIAA), the National Space Society (NSS), and the National Aeronautics and Space Administration (NASA) for providing a wealth of pictorial and scientific information on their websites.
Clearly nothing would be possible without God in my life who blessed me with wonderful parents Walter and Dorothy Clements. Of course, I need to thank my kids Austin Clements and his wife Crystal and my daughter attorney Malaika and her husband Ken Billups for the support and editing comments with solid suggestions. And most of all, I thank, the love of life, my lovely wife Eugenia Davis Clements for her invaluable comments, relentless editing, steadfast support and encouragement throughout the entire process. Again, thank you all.

INTRODUCTION

This story is a saga that will likely provoke deeper questions and thought. Elements of surprise, drama, catastrophe and spirituality are all enshrined by scientific principles.

Many of us enjoy a placid existence despite the dominating problems of the twenty first century such as staggering problems of war, gang violence, global warming, terrorists, school shootings, etc. Imagine what it would take for all of these problems to just go away? What uniting force could make us all work together for a common cause for the human race?

This conceivable situation requires us to put aside differences and work together to overcome a challenge never before conceived. Hopefully the motivation presented in this story will force us to think out of the box and open our eyes to possibilities never before imagined.

The story opens with inexplicable events that are somewhat alarming and invoke a watchful concern. As mysterious events escalate greater attention is focused on them. The key characters are in a unique position to help identify the escalating calamity if only allowed to use their knowledge and skills. Fear continues to intensify and chaos and pandemonium ensue forcing humanity to focus on the dilemma at hand. The astonishing nature of the problem is finally revealed but the solution is equally mind-boggling. Mankind is compelled to work together if the solution is to be successful.

CONTENTS

1

GPS Satellites Vanish

An astronomer peered through the deep space telescope located in the remote, sprawling high-tech facility of the Paranal Observatory in Chile, operated by the European Southern Observatory. The monitors allowed him to scan with the world's most advanced visible-light astronomical observatory. He looked at the New Events menu on the monitor and noticed a strange blinking. The dimly lit object grew larger and clearer, indicating that something was streaking toward Earth.

He turned his head to his colleague. "Hey, Clarence! You need to come take a look at this. Quickly! Something really weird is showing on the New Events screen."

Before Clarence budged, the object vanished. Suddenly, a jolt surged throughout their third-story room, nearly knocking them to the floor.

"Jesus! What in the hell is going on?" the astronomer shouted.

"What are you talking about?" Clarence said. "The object or the jolt?"

"Both! Was that an earthquake? And I don't know what in the world I saw on the monitor, but it's gone."

"It did feel like an earthquake. But I have no idea what you think you saw on the monitor."

"It came toward us and was gone in a split-second. Then this whole building rocked. We have to do something *now!*"

"Calm down, man. It was just an earthquake. We've seen worse."

"Yeah, but I have never seen anything like what I saw on the screen."

"Maybe it was a meteorite or asteroid on something."

"No. I've seen those many times. This was something else."

At the same time in the United States at the Colorado National Geospatial-Intelligence Agency (NGA) several scientists were tracking GPS satellites. These US mission specialists were tracking the flight paths of their designated GPS satellites from the thirty-two different GPS satellites orbiting Earth.

One of the mission specialists peered at his monitor and witnessed a sudden flash across the screen as two satellites abruptly vanished. An instant later, the entire room jerked and rocked, and the bank of alarms at the ground-station facility sounded. Monitors flashed, showing that all communications for two GPS llF satellites had been interrupted and they had both vanished from the radar screen. The team scrambled to figure out what had happened.

"What in the world is going on?" one of them shouted. "That felt like an earthquake!"

"Impossible! We're on an isolation slab. We're immune to earthquakes."

The Internet and phone communication then cut off, causing total confusion in the room. "Forget about an earthquake. It can't be that," another specialist said. "But did that flashing object just knock out two of our satellites?"

"What object? We haven't detected any object. What are you talking about?"

"But I saw something! Let me play it back."

As the alarms continued to sound, others outside the room ran down the hall in a state of panic, shouting, "What's going on?"

The scientist in the room gathered around the mission specialist's monitor to watch the replay.

"Look at that flash. I told you! What the hell was that? See? The two GPS satellite signals just vanished. Disappeared!" the mission specialist said.

"Beats the hell out of me," one of his colleagues responded. "Something must have knocked them out."

"There's no way we could have missed something big and fast enough to wipe them out."

"What about a meteor or some type of asteroid? Or maybe a terrorist missile?"

"Come on, man! Don't be stupid. We would have detected something like that. There was no explosion, no debris—nothing."

"Well, for one of our satellites to have vanished, let alone two, something else had to have happened or something else is out there.

2

Classroom Lecture Disruption

A few minutes earlier, Jason Scott sat in a UCLA classroom in Westwood, California. He struggled to keep pace with the notes that Professor Duncan Thatcher, the visiting lecturer from Harvard University, was scribbling all over the classroom board.

Only twelve graduate students were in the classroom that seated more than forty. The students clustered in the front. The sunlight peeked through the mini blinds, reflecting off Thatcher's balding head as he stood with his back to the class, displaying his unkempt, dumpy body. He had bone-white curly hair crowning his shiny head and somewhat villainous facial expressions. His Coke-bottle, black-rimmed glasses magnified his brown eyes, making them look comically enormous. The professor's thick British accent didn't help matters for understanding his pedantic discourse.

Jason gritted his teeth and whispered to his long-time best buddy, Austin, "When is this guy going to cover our stuff?"

"I don't know bro. He's adding nothing of value to our research. He keeps interjecting all this irrelevant environmental stuff," Austin whispered back.

"Look, we can suck it up, or confront him again."

Professor Thatcher had promised to cover the dynamics of interstellar winds and moving bodies. Jason and Austin's PhD team needed the information in order to stop relying on the cosmic wind algorithm Thatcher had developed otherwise they would have to start from scratch and would run out of time. Their research would be doomed if they couldn't add that last piece of information.

"Can't confront" a whisper blew from Jason to his friend. The last time there'd been a confrontation with Thatcher, Jason had been highly embarrassed and humiliated. Thatcher had labeled him as impatient, arrogant, and spoiled. Jason squirmed in his seat.

Jason and Austin believed their Harvard competitors were moving forward with their research while Thatcher was literally blocking their progress. They knew that if the Harvard team published their preliminary paper first, the UCLA team's work would be null and void. They had been working on this breakthrough for too long, and were on the verge of something really big. "We can't continue being polite and letting this jerk get in our way now," one of them whispered to loud.

Professor Thatcher, hearing some mumbling, turned around and saw Jason whispering to Austin. "Excuse me, gentlemen! Would you like to come up here and share your important conversation with the rest of the class?"

Jason looked at Austin and wanted to reply in a strong, *colorful* manner. Instead, he took a deep breath, looked down at the floor, and managed to bite his tongue. Austin was not so restrained.

He flipped his long brown dreadlocks over his right shoulder, revealing his gold chain necklace. He had a handsome ebony face with striking hazel eyes. In his standard sleeveless, V-neck sweatshirt, he clasped his hands, shrugged, and then crossed his arms, exposing his muscular biceps.

"Sir, to be frank, this whole lecture has been a waste of time for us. You haven't covered the key topic you put in the course outline," Austin blurted before Jason got the chance to speak.

Insulted and incensed, Thatcher shot back, "You *government Negros* are allowed to be in this school, and yet, you still don't appreciate the opportunity to be here in my presence."

Lurching from his seat, Jason stepped toward Thatcher. Austin jumped up and grabbed Jason before he overreacted.

Thatcher, startled, inched backward, stumbling over the wastebasket behind him. He gyrated to keep from falling but hit his hand on the desk before toppling to the floor. He landed on his back with his legs flailing in the air.

Jason and Austin rushed to help him up.

"Are you okay, sir?" Jason said.

"Get your hands off me!" Thatcher wobbled to his feet, holding his wrist.

Jason and Austin backed away towards their seats.

"Sir, I hope you're okay. I wasn't going to attack you. But your comment to my friend was uncalled for. Austin can run circles around most students. He only told you what all of us on the Galactic research project were thinking. You *have* been wasting our time." Jason looked around at the stunned students and the fuming Thatcher. "Sir, you need to apologize for that remark. But I suggest we reset and refocus on the lecture from here."

Jason sat back down, hoping everyone would stop staring at him and things would settle down. He flipped his black, wavy hair, tied back into a ponytail, over his neck and shoulder. His long, athletic legs stretched into the aisle where his well-worn, black high-top tennis shoes jutted out from his cuffed blue jeans. As always, the sleeves on his sports shirt were rolled up to three-quarters length, making him look more businesslike. Sometimes, his buddies called him Tarzan because of his resemblance to the action hero's looks and similar conduct.

Austin, still quietly fuming from Thatcher's insult, concealed his smart phone and sent a text message to his homeys. "Let's hook Thatcher this evening. The dude just trashed us government people and claimed we should be happy to kiss his ass."

Professor Thatcher stood up and stared at Jason. Thatcher wore his years of formal research in astrophysics at Harvard and Oxford on his sleeve. Back in the day, he grappled with such dense concepts as planetary gas storms, galactic winds, solar wind phenomena, and solar plasma propagation.

He had a narrow area of expertise—mapping the subtle but cumulative dynamic effects of galactic plasma winds on giant stars. He had thought this would be a major analytical tool. He had hoped to claim it as his own to catapult himself into the national spotlight. But he had never documented the project as it had grown more complicated—a tendency that had plagued him throughout his entire career.

His obsession, more than a dream, was to receive credit for a breakthrough in astrophysics, and he had focused on being in the right place in case an opportunity emerged. His motivation for coming to UCLA had been to join one of the top teams doing breakthrough research on the cosmos. And a cosmic wind research team was one of two such teams at

UCLA. Jason's galactic body dynamics team was another similar but competing team.

Jason's classmates, still stunned by Thatcher's words and his tumble, watched as Thatcher attempted to regain his composure. The room was silent.

Jason peered around at the shocked expressions. *Oh no,* he thought. *What the hell did I just do to this vengeful, arrogant man?* Beads of sweat formed on his forehead.

The professor stood silent for several moments with eyes looming large behind his thick lenses. Finally, he lashed out, "Now what about you other students?" He switched his gaze to the rest of the class. "Does anyone else feel the same about my lectures?"

The classroom was quiet. *Do they really want me to change the format of my lectures? I'll be damned if I'm going to go out of my way to please these jerks,* Thatcher thought.

Within a few minutes, other students began voicing similar complaints. Listening attentively, Thatcher looked around the classroom and then fixed his gaze again on Jason, who was looking down at the floor. Thatcher's look was filled with disdain.

"Well, thank you, folks, for your candid comments. I will digest them and decide on the appropriate action," he said. *I'll fix these smartass bastards.*

Thatcher had been motivated to leave Harvard to pursue his need for recognition elsewhere but he was critical of UCLA. He once said, "UCLA has a strong athletic program and is recognized for that worldwide, whereas Harvard has a weaker athletic program but is recognized worldwide for its academic excellence. Students have a choice—brawn or brains."

3

Earthquake Jolts Classroom

As Professor Thatcher walked toward Jason to chastise him further, a sudden jolt shook the entire room. The walls undulated, and the window blinds rattled. Everything in the room swayed, and the students instinctively dropped to the floor. They could hear others screaming outside the classroom. Professor Thatcher let out a pig like squeal and crawled under his desk.

A sonic boom roared somewhere in the distance. The ceiling creaked, and the building continued to sway. Some students crawled under their desks while others scooted away from the windows for fear the glass would implode at any moment. Professor Thatcher remained frozen under his desk, curled into the fetal position and squealed like a pig.

"Earthquake!" someone shouted.

Voices just outside the doorway yelled to one another. "Don't panic! Stand under an archway in case something collapses!"

In the courtyard, students scrambled to get away from the buildings surrounding them, hesitant to seek cover for fear the structures might collapse.

Back in the classroom, the roaring noise that accompanied the quake seemed to last an eternity. Eventually, the creaking sounds dissipated.

"Good grief!" Jason exclaimed. "Is everyone all right?"

"Man, that was awesome!" beamed Austin. "That was some serious shaking, dude!"

No major physical damage to the room was apparent, and everyone sighed with relief. Professor Thatcher peeked out from under his desk in complete disarray, his spectacles hanging onto his head by one ear.

"Are you sure that was an earthquake?" one student asked.

"You're damned right!" another student said. "It swayed more than anything I've ever experienced. It was probably massive at the epicenter. You know the average earthquake lasts about sixteen seconds. This one was much longer. But don't ask me what that weird sonic boom was all about."

Professor Thatcher emerged slowly from under his desk, disoriented and still frightened. Trying to regain his composure, he stumbled back to his place at the front of the room while the students returned to their seats. Although he remained front and center, he was far from fine. Having observed his unmanly response to the quake, many of the students were restraining their laughter. Humiliated, he barked, "Another one of those jolts and I will leave you little brats and get the hell out of this town! I will pack up and go back to Harvard!"

"This guy sounds and thinks like a Piglet," Jason smirked and said under his breath as he turned around but his words were a little too loud.

Many snickered.

"There *will* probably be aftershocks, Sir. So you may want to get out your luggage," Austin said.

Thatcher glared at Austin. "Give me a tornado or a hurricane any day." "You kids can have these damned earthquakes."

Jason tried to restore some semblance of dignity to the classroom and to the professor. "I'm sure we will hear more about it on the news tonight. It probably caused widespread damage."

"You never know how severe an earthquake really is until you know the location of the actual epicenter," another student interjected. "If it were nearby, we'd have already observed the majority of the damage. However, if the epicenter were far away, like in Palm Springs, considering what we felt here, this temblor could break records."

"You may be right," Professor Thatcher said. "Anyway, let's finish the lecture. I think we'll be alright."

Regardless of the professor's remarks, several of the students knew that the possibility of aftershocks remained high, so their safety was far from assured.

As Professor Thatcher struggled to get his lecture back on track, he noticed two students staring out the window.

"Excuse me!" he said. "Please have the courtesy to pay attention to what I am saying!"

"Professor, you need to see this," one of the students directed.

"See what?" Professor Thatcher asked.

The students pointed toward the sky. "Really, sir," one student said. "You should come over here right now."

The professor and several students walked to the window and looked outside. Spanning the entire sky appeared what looked like three faint, eerie skywriting bands or scrolls, except wider and flatter. They were light brown and had something that resembled scribbling inside of them. They churned and glowed and were the most bizarre things any of them had ever seen.

"I've seen skywriting before. This may be some strange environmental disturbance," Professor Thatcher said.

"Yeah, man, it's really weird looking," Austin agreed. "It's probably some type of missile or rocket contrail. We need to check out what the military is up to now."

The other students, equally puzzled, began speculating.

"I don't think any of us know what it is," Professor Thatcher said. "Let's get back to work. We have had too many disruptions already. As one of you said before, they will likely tell us all about it on the news tonight."

"I hope they do. Something is very strange," Jason pensively figured. "I can't put my finger on it, but for some reason I have a feeling that this *was not* just an ordinary earthquake.

4

The Tree Topples

Professor Thatcher resumed the lecture despite the debris on the floor, thinking this gutsy move might erase the students' perception of his earlier behavior. He discussed the notion of black holes and supernovas in conjunction with the emerging notions of membrane and string theories of the universe. But this still did not help Jason and Austin with the missing gap in their research.

Suddenly, there was a loud *crack* just outside the building. Again, the professor let out another high-pitched squeal and darted under his desk. Everyone jumped and jerked around to focus through the mini blinds. A large tree was slowly falling toward them, and an eerie irregular shadowy figure was plunging toward the windows. Before anyone could move, the tree slammed into the windows, and everyone fell backward and cringed. The ceiling light fixture snapped open, and one of the florescent tubes fell to the floor, shattering. Fortunately, the exterior burglar bars protected the windows. The rails were bent and jarred loose from the outer wall, but none of the windows were broken.

Several students ran to the windows. The base of the venerable tree was fractured through to the roots.

"Big Brutus fell!" one student yelled.

"Now that's what you call a big-ass angry tree," Austin blurted.

Trembling and freshly disheveled as he crawled back out from beneath his desk, Professor Thatcher commented, "Thank God that tree didn't break the windows,…. Okay, I think we have had enough for one day. Class is dismissed. I'll see you later. I'm out of here!"

Jason and Austin walked outside to survey the damage. The cluttered courtyard surrounded by towering buildings showed no major structural damage.

Some students milled around while a few others examined and bemoaned the fallen 400 year old Brutus.

5

Friends Past Encounters

Taking a look back to the past, Jason and Austin met in high school in a science class. Austin was the charismatic, notorious skirt chaser. He would swagger down the halls and make some sort of flirty remark toward any girl who caught his attention. Many of the girls felt pestered by him while others loved the attention.

"What are you going to do with all those phone numbers you collect from just about every chick you meet?" Jason said.

"Look-a-here, dude. I don't know. I just want to take'em and make'em," Austin winked and snickered.

"You're heading down the path to party away your future. Are you going to work with me on our science project, or are you going to follow every girl who passes by?" With eyes piercing straight into Austin's, Jason continued "If so, we need to part ways now,".

"Oh, c'mon, dude! We can do both. Just chill a little bit. I can hook your lanky, clumsy, track-and-field ass up with something you can't pass up," Austin offered.

"Don't worry about me, dude. I can handle myself," Jason said. "You need to decide which path you want to take. You have an amazing mind and awesome academic skills, man. I have a lot of benefits because of my parents though I *do* have to accomplish things on my own. But you, you can run circles around most of the students in our class."

"Thanks, bro. I *am* rather cool and sharp," Austin agreed, quite self satisfied.

"Austin, you have to decide if you want to work hard with me on this project or chase skirts all day long. We can kick butt, but it's up to you, man."

Austin and Jason couldn't have come from more different backgrounds: Austin's ghetto life versus Jason's highly cultured, upper-middle class with both parents working for NASA. Austin's mother died when he was just ten, leaving his father alone to raise four hard-headed boys. As a result, Austin developed a fear of abandonment by women. .

Austin turned and started to walk away from Jason but he stopped in his tracks and stood contemplating for a moment. Finally, he turned back and giving in said, "Let's kick butt."

After that little episode, Jason took Austin under his wing, and Austin was grateful to Jason for helping him see beyond some of his immature limitations. Eventually, Jason and Austin were almost like brothers. They were committed to and cared for each other, and together they overcame many adversities. Their classmates often teased them after watching 'the black guy and white guy' work closely together on science projects after school while sometimes skipping track practice. Their award-winning high-school science project, in which they designed a space station to orbit the moon, was fast becoming a reality as their education continued to progress.

Back at the UCLA campus, Jason and Austin saw Brutus lying outside their classroom.

Friends Past Encounters

"Man, look at all the branches strewn all over the courtyard," Austin said.

"And look at all the overturned trashed cans along the sidewalks," Jason said. "We gotta find Jane. My cell phone seems to be out. I hope she's on her way to the cafeteria."

As Jason and Austin hurried toward the cafeteria, they felt the uneasy mood on campus. Many students continued looking skyward and pointing toward the scrolls, nervous and bewildered at the sight.

To their relief, Jason and Austin saw Jane waiting for them in front of the cafeteria. Her long, wavy hair was drawn back in a ponytail and gently wavered in the gusty breeze. Her hair was much like Jason's. In fact, looking at them from behind, one could hardly tell them apart. Austin referred to them as Tarzan and Jane and would sometimes follow this up by bellowing the Tarzan yell. Jane waved at Austin and ran to Jason, giving him a big hug.

Tall and shapely, Jane's movements and mannerisms flowed with grace. She dressed conservatively, preferring to wear oversized hooded sweatshirts and loose jeans that concealed her gorgeous body to ward off flirts. Unlike Jason, she had endured many hardships, especially when she and her parents had to take care of her infirmed grandmother for a number of years. This probably gave rise to her conservative views and faithful church attendance. Jane had simple ambitions and goals: get married, have children, and perhaps open her own antique shop.

Jason, Austin, and Jane liked to have lunch in the same spot where many of their friends joined them to chitchat and argue. The south cafeteria was the largest of several that were scattered throughout the beautifully manicured lawns of the sprawling campus. The towering green trees imparted a sense of tranquility. The shaded walkways remained clean for the most part and free of clutter. But not this afternoon.

The students loved the cafeteria's modern and accommodating design. TVs hung from the ceiling in several sections while other areas were secluded for those wishing to study. Still other areas had elevated sections for panoramic viewing—a feature that attracted many males who passed the time watching the females who strutted by. The food venues ran the gamut from pizzas, hamburgers, and hot dogs to full meals, desserts and salad bars. The main course meals were rated "simply awesome" by the students.

Jason, Jane, and Austin remained unnerved when they saw the disheveled chairs and tables through the windows in the cafeteria.

"Hey, I don't know about you lovebirds, but I'm hungry as hell," Austin said. "I need to get some grub before I get a headache from all this turmoil."

"I agree. Jane, let's go with Austin," Jason said.

They entered the food lines, surrounded by frenzied students talking about the earthquake and the mysterious scrolls in the sky. Jason emerged from the line first and made his way to their favorite eating spot. Jane's line stalled for a few moments as a student argued about the price of his food.

"Go ahead, Tarzan!" Austin gestured for Jason to speed it up before someone else grabbed their spot, as none of their other buddies were there yet. They were still outside, staring up at the sky scrolls and canvassing the campus for earthquake damage.

Friends Past Encounters

Jane gripped her food tray with both hands and ran to catch up with Jason, cutting in front of Austin. He allowed her to get in front of him but whacked her firmly and loudly on the butt. Jane whirled around and noticed Austin chuckling at her look of surprise. She grabbed the water cup from her tray and hurled it at him. He managed to jump out of the way and the water splattered on the floor.

"I should kick you in the balls for doing that!" Jane blurted out.

"Ouch! What did you say?" Austin was stunned to hear such language from the normally polite and conservative Jane.

"You heard me, you jerk!"

"Damn, you are way too sensitive," he said. Then he added slyly, "I'll let you kick me there if you take off your shoes and be gentle with your little toezies."

"Austin, you have such a dirty mind." Jane frowned. "No wonder you have so many different girlfriends."

"You mean that is why I am *able* to have so many girlfriends."

"Oh, shut up! You really do have a one-track mind."

"Anyway, if you hurt me, I may not be able to have kids. So will you then allow me to be the godfather to Jason's and your child?"

Jane and Austin's squabbles were nothing new. They were always competing for Jason's time and friendship.

Jane's high moral standing was the polar opposite of Austin's corrupted demeanor. She believed in the sanctity of marriage and religion and was an adamant pro-life supporter. Austin, on the other hand, had no qualms about screwing any willing young lady if the opportunity presented itself. He once advised Jason that he should not count on women being there when you need them because when the going gets tough, they would leave you standing alone. This generalized philosophy was borne from the early death of Austin's mother. This apparently instilled in him and his brothers a strong mistrust of women and the lack of respectable, good home training promoted episodes of family disrespect.

On the other hand, Jane and Austin did agree on one thing about their mutual friend, Jason. They recognized his inner conflict between his religious upbringing and his intense scientific yearning, which they felt jeopardized his belief in God.

One defining event forever changed Jason's values. His parents were very proud when he graduated from college. But they unknowingly created a difficult situation for Jason when they asked him to speak to some church folks when he arrived home from college. Jason agreed, thinking it would only be a few of the elderly people. But his parents managed to invite two congregations of a few hundred people and get their son listed as one of the keynote speakers.

When shy Jason arrived and was escorted to the front table, where he saw his name on the program, he almost passed out. He was so nervous and traumatized that when they finally called him to the podium, he freaked out. He stood up and calmly strolled right past the podium and out the side door. His parents were mortified.

From that day forward, Jason resented the institution. He went through the motions of going to church, but his belief was shaky at best.

The three of them finally settled down in the cafeteria at their usual spot. Jason eventually described the incident with Thatcher stumbling over the wastebasket and injuring his hand in class. Jane often worried that Jason would lose his temper and erupt in class.

"Jason, you can get kicked out of the class or even school for behaving like that," Jane said.

"I was just so damned frustrated from wasting so much energy on the wrong thing, but he needed his butt kicked for his comment about Austin. Thatcher was lucky he fell first."

"That was stupid. We all get frustrated, but you should realize you can't always have things go your way—"

"Look, Jane, I'm here to learn and complete my research project. I'm not here to practice note-taking. We have been working our butts off for years on our research, and we took his class for one reason only: to help us with one piece of technology to complete our research."

"I understand that, but why didn't you simply explain your opinions to him and not lose your cool?"

"We have on several occasions."

"C'mon. The professor is a jerk," Austin interjected. "All of us feel the same frustration as you, bro."

"Are you going to apologize for your outburst?" asked Jane.

"Why should I? We're really going to pressure him to start covering the missing technology."

"Maybe you should just say you're sorry," Jane suggested. "It may defuse the situation."

"Hell no! I am simply trying to learn what I need to help with my research. I don't have to be diplomatic and kiss ass. They can't do anything to me anyway."

Jane shook her head and slapped the table. "Don't be so cocky. This could have derailed your celestial body research project since you are now heading it up."

"I don't think so, Jane. Professor Thatcher doesn't have any authority there. Professor Stein is the chief scientist heading the research."

"Dumbo Thatcher can't hold a candle to Professor Stein," Austin said. "Professor Stein is supposed to have an IQ of 165. But he acts like an absented minded professor at times. The dude forgets to put the cap on his pen. That's why you see those ink stains on his shirt pockets."

"Like all my professors, Stein probably works very hard to make himself look good here at UCLA," Jane asserted.

"Not really, my sistah," Austin said. "Stein is not just looking out for himself. He will watch your back too. The dude focuses on his research and does not give a damn about politics and administrative stuff. He would rather see his students learn and prosper from his efforts rather than receive any recognition himself. We believe his greatest aspiration is to propel at least one of his students to receive national scientific recognition. Thus, we have Jason."

"You're right, Austin. I only hear good things about Stein," Jane acknowledged.

"Thatcher, with his Harvard background, thinks he's God's gift to academia," Austin continued. "Although I heard he left before wrapping up his research there. Anyway, most of the times you can't even understand him because of that annoying British accent."

Friends Past Encounters

"You're right," Jason said. "I may not even mention the incident to Professor Stein, but he would surely side with me. Anyway, he already knows what a jerk Professor Thatcher can be."

"I don't want to keep harping on this, but do you think your outburst might derail your chances of being nominated for the award?" Jane said.

"You mean the Nobel Prize nomination?"

"Yes."

"I don't know why it would," Austin said. "The lab secretary told me they have been bugging Professor Stein. He was supposed to meet with the engineering dean to supply him with administrative information about Jason and our project. She sounded sure this information gathering was being done because the school wanted to submit our breakthrough research work on the physics of celestial bodies for the nomination. Jane, do you know about the talk going around the engineering department?"

"What talk?"

"Well, the entire technical community claims our research is leading the pack in the astrophysics community, even over Harvard," Austin said. "My man Jason could be the youngest recipient ever nominated. Usually they nominate the professor, but Jason's breakthrough contribution has surpassed all of them."

"Come on, Austin. You're a major contributor. When we get through all of this, we'll start working on our lunar space station," Jason said.

"Stein has been working night and day to make sure your nomination package is superb and secure . That's why it's late. He also believes you are the one. So watch out, my brotha and sistah. I'm going to be promoting Tarzan and Jane around here one day to save the world."

"Yeah, yeah, yeah," Jason grumbled. "If I get nominated, that's great. But I am not going around kissing ass just to get nominated. I am only going along with this because it's Professor Stein's dream. Besides, this nomination is just the first step in a long, drawn-out process. What's most important to me is being able to understand scientific phenomena and solve scientific problems and mysteries."

"Well, that's a great and noble goal, my brotha," Austin pointed out. "I wish you would also work with Mother Nature and have her take it easy on us with these unpredictable earthquakes."

"I don't think she has anything to do with it," Jason said with a puzzled look.

"What are you saying?" Jane asked.

"I don't think Mother Nature is behind all of this."

"What the hell are you getting at?" Austin said. "I think Mother Nature has every right to be pissed off at all the shitty things we are doing to her. Destroying the ozone layer and the tropical rain forests and sending all of this hardware debris into space."

"Austin, don't start with your environmental whining."

"Say, bro, you had better start listening to that whining. I read yesterday where NASA now tracks ten thousand pieces of space debris, all bigger than a softball and still orbiting the earth. Space pollution is growing every day. Further, at any given time, there are over three thousand airplanes in US airspace, each one spewing burnt jet fuel into the atmosphere. I think our Maker may finally be getting even with us."

Friends Past Encounters

Jason shook his head and frowned. "Like I said, Mother Nature may not be the problem anyway. I don't know what's going on, but we need to deal more with real science to gain answers and not worry about disturbing some rain or sun god."

"Okay, boys, we'll wait for science to come to our rescue," Jane said, trying to defuse the escalating argument. "By the way, since your dad is on the board of the alumni association, don't you think you should tell him about the incident in class so he won't be blindsided?"

"Oh, damn, that's right!" Jason slapped the table. "He'll come unglued if he finds out about the stupid Thatcher tumble. My sister and I are going to call him tonight at the Cape, so perhaps I'll bring it up. It could stress him out. NASA frowns on family members acting up just before a space station launch. I'll have to wait and see. Britney and I will focus on all the other stuff that happened today, like the earthquake and that weird stuff in the sky."

As they stood to leave, a female student carrying her tray of food slipped on the water Jane had left on the floor. No bus boy had wiped it up. The girl made a desperate but failed attempt to prevent her fall. In doing so, she flung her tray and banged her elbow on the table but took a vicious fall, slamming her head on the floor. Several of the students sitting nearby ran over to help.

Jane had her back to the incident and did not witness the fall. When she heard all the commotion, she started to walk over to see what was happening. Austin grabbed her arm.

"She'll be okay, Jane," Austin said. "Jason and I need to get out of here now. We're already late."

Jane started to pull away from Austin, but he managed to coax her to leave, and she reluctantly exited the cafeteria with her friends.

One of the male friends of the fallen girl, angered at Jane's apparent lack of feeling, started to dash over to Jane, but his buddy grabbed him and said, "Just let her go. I know where that uppity chick lives."

6

The Professor Seeks Revenge

Professor Thatcher headed toward his research lab and noticed the overhead sky streaks. Suddenly there was a bright lightning strike and a thunderous clap in the vicinity. Several nearby students cringed and ducked.

"What the hell was that? Lightning from nowhere… in the middle of the morning. No clouds anywhere except those creepy cloud streaks. What's going on?" one student blurted.

Thatcher trotted ahead and lumbered into his environmental global warming research lab and saw several of his key students working on their computers and chatting about the earthquake and sonic boom. They had also noticed the sky scrolls, and noticed the professor's disheveled appearance and the alarm on his face.

"Is everything okay, Professor Thatcher?" one of his students in a competitive group of Jason's said.

Thatcher laid his briefcase on the table. "I can't believe the nerve of those cosmic research students. They are incredibly arrogant, and Jason is downright dangerous and called me a pig," he grumbled.

The seven students turned to look at Thatcher. Lamont's eyes shined as he pointed his finger at Thatcher.

"Sir, I'm not supprise that happened with Jason?" Lamont said. "You know we all have had trouble with that arrogant jerk. I owe that asshole a payback for trying to put me down in front of my girlfriend when I tried to enlighten her about man's impact on the environment."

"Don't worry about it now," Thatcher said. "I'll take care of that guy later. You students keep up the good work on our planetary global environmental models. We won't let those goons get ahead of us."

7

Space Reunion

That evening, Jason and his twin sister, Britney, who wore a lip ring and had a tattoo on her neck, were preparing to place a closed-circuit conference call from their home to their father, Walter, in Florida at the Cape Kennedy launch site final-stage holding area for the manned SpaceX's Falcon 15.

"Do you feel special that NASA has set up these funky communication lines for all family members of the astronauts to chat just before a launch?" Britney said. "I think that since Daddy is the lead astronaut for the incoming exchange team with the International Space Station, we are entitled to these cool privileges."

Puzzled, Jason mumbled, "What privileges? All the families are treated fairly, my dear, so that each astronaut feels equally valuable on the little space station."

"I think I could be comfortable up there," Britney said. "The space station is the size of two Boeing 747 jetliners. I understand that with the solar panels and support structures, it is nearly the size of two football fields. It weighs 470 tons and carries a crew of seven astronauts. It orbits the earth at 217 to 285 miles and whizzes by at 17,500 miles per hour, completing one orbit every ninety minutes. That's not too shabby, big brother."

"I'm impressed that you seem to know more about the mission than I thought. But do you actually know anything about the mission, little sister?" Jason questioned.

"Well, in layman's terms so you can understand, they may call Mommy Captain Dorothy up in the space station, but she will be up there for three and a half months doing assembly work on the environmental biochamber and serve as one of the communications officers. The new bio-chamber is a bus-sized module designed to provide self-contained environments using plants and recycled waste. It allows astronauts to live in space for extended periods of time."

"I'm impressed that you paid attention and memorized all that technical stuff," Jason said.

Their father planned to make his second flight into space on the Falcon 15 and join their mother, Dorothy. Together they would realize their lifelong goal of jointly operating the International Space Station. This would be the last time Jason and Britney could talk to their father before the launch scheduled for the following morning.

"I'm looking forward to chatting with Daddy, but I know I'll get bored when you guys start talking all that technical stuff," Britney said. "I have faith that those NASA nerds know what they are doing, and that Mommy and Daddy will be safe throughout the mission. Otherwise, I'll have to hurt them. So I don't think you guys have to debate the agency's procedures and protocols again and again."

"So, lady, what do you want to limit our talk to? Rock music?"

"I'm tired of pulling these all-nighters to keep up," Britney claimed. She was a student at Cal State University at Long Beach (CSULB), pursuing a degree in public relations. "We have opposite personalities. I don't have self-confidence in science partly because everyone in this family is so science and technology oriented. Remember, in high school I resisted the science career path when you guys put too many expectations on me. I want my career choice to be intuitive rather than follow in the footsteps of you, Mom, and Dad. You guys still have well-defined, short- and long-term fixed goals. So what if I'm more impulsive and likely to overreact to stress or uncomfortable situations?"

"Sounds like you're ashamed of yourself," Jason charged.

"No, I'm not. I like myself. I like my body piercings and all of the jewelry that I wear," Britney stated. "Furthermore, I have a awesome singing voice that I'm thankful mom and dad helped nurture."

"Yeah, by spending all that money on expensive voice lessons for classical singing," Jason said.

"I've graduated from all that old-fashioned classical music to modern sounds."

"Modern sounds? You call your rock band 'modern'? You call yourselves The Metal Rocks, right?"

"The band members want me to join them. They believe they're about to get their big break to go on tour."

"You have got to be kidding! Mom would have a fit if you did that!"

"That's why I wouldn't even dare tell Daddy. I once mentioned the group to Mom, but not as a serious pursuit."

"I think you have gone off the deep end, Britney."

"Jason, you have a single-minded focus on goals and an intense determination to achieve them, which gives rise to your greatest shortcoming: you are a closet agnostic and believe that the nature of God is unknowable."

"What are you talking about?"

"You believe people are in control of their own fate." Britney walked around the table to confront Jason. "You're insensitive to those who claim cultural hardships and obstacles prevent them from achieving their dreams. I think it gives you a distorted view of life's real trials and tribulations."

"What brought on this sermon?" Jason inquired. "Come on, Britney. So you believe I have a distorted view of life?"

"Jason, you often argue against claims that environmental and cultural factors have a limited effect on one's success, that success is primarily based on hard work, determination, and overcoming adversity. That's just the way you think."

"Whatever Britney. You are way off base. But let's call Dad now."

They called using the speakerphone system on the dining room table. When they finally linked to their dad, they brought up the earthquake and how it had shaken up everyone at both campuses. Jason remained baffled that such a large earthquake caused so little structural damage. Their father had heard that some NASA officials were scurrying around the site investigating unusual ground motion, but he was certain there was no relationship. The launch vehicle was on an isolation slab and immune from the ground motion. The officials had also heard about the lost communication of two of the GPS satellites but were not allowing the astronauts to be concerned.

Space Reunion

"After we talk, I'll switch to the late night news," Jason said. "They may have more details."

Despite reservations to bring up personal events, Jason mustered up the courage to mention Thatcher's tumbling incident during class. Britney was shocked, and their father expressed extreme disappointment in Jason for venting his frustrations in such a childish manner.

"Jason, you have too much at stake to behave like that," Walter said. "You have to stop reacting so impulsively. Your mom and I let you two have your way too often, and now it is backfiring. It looks like neither one of you two will be part of the new space frontier."

Jason and Britney listened to their father's reprimand. Britney was not surprised to find herself once again included in her brother's misbehavior.

"You should have calmly approached the professor with some of your classmates after class," Walter continued. "Then you should have discussed your displeasure collectively and intelligently. You could have calmly suggested possible changes in the class lecture format."

Walter paused for a moment, and there was silence among the three of them. Britney mocked her father's reprimand by mimicking conducting Beethoven's Fifth.

"Anyway, Jason, I hope you are strong enough to handle this and accept the consequences you may have inflicted upon yourself," Walter added.

Britney gestured in the background with a clenched fist. She was thrilled to hear about her spoiled brother's unbecoming, threatening exhibit to Thatcher. She gave Jason a high-five, her large, glitzy earrings jingling as she hummed a soft, melodic tune. Jason gestured for her to keep the noise down as their father continued his lecture on courtesy and demeanor.

"You are so spoiled," she whispered to Jason. "One day things are going to backfire on your spoiled ass."

"I have some bad news," Walter announced. "You know NASA has been reviewing and upgrading its personnel policies related to spaceflight and space missions. They intend to implement a new safety policy in the future forbidding both parents from being on the same space mission at the same time. This is intended to protect a family should a catastrophic event occur during the mission."

Britney was beside herself. "You must be kidding! Does that mean you and Mom won't be able to continue to work in space together?"

"That could happen."

"Daddy, that's ridiculous! You guys have dreamed about upgrading the new space station together."

"That's true. I guess we will have to change our plans. Anyway, we may now have to focus on other projects."

"But you guys planned this well before Jason and I were even born! I can't believe those NASA people."

"Britney, listen. They are only trying to protect the children."

"Daddy, we realize there are risks."

"Well, then, can't you see that it would be better if only one parent were lost rather than both?"

"This policy is simply stupid. You shouldn't allow those buffoons to void all of Mom's and your plans just like that."

"Watch your mouth, young lady," Walter said. "They are trying to do what they feel is best."

Space Reunion

"Good grief, Daddy, you're right! People should do what they feel is best. I'm going to quit school and go on tour with The Metal Rocks!"

Jason's jaw dropped, and he immediately pressed the mute button. "What the hell did you say, Britney? Quit school and join that band? Are you crazy?"

"You heard right."

"Britney, just shut up. Let's talk about it later. Don't let Dad know about this. He'll go ballistic in the Falcon 15. He may blow the whole NASA mission!"

Britney crossed her arms and pouted. Jason pressed the mute button again just as Walter asked, for the second time, "Did you say you're quitting school?"

Jason reacted quickly. "No, Dad. She said she wants to be in a *quick* school so she could graduate sooner." Jason quickly changed the subject. "Dad, you and Mom are special. You are the only astronaut parents in NASA."

Walter hesitated for a moment. "No, that's not true. Don't forget about Larry and Bennie. And Larry and I are on this same launch tomorrow morning."

"That doesn't count. Bennie is retired. She stays at home and lives like a celebrity."

"I would not be too quick to call her a celebrity. By the way, Larry and I were just recalling the time when our families vacationed together at Disney World. Do you guys remember that?"

"What about it?"

"You remember when Mom, Larry, Bennie, and I were with all of you kids and you could not wait to get into the Space Station exhibit at the theme park?"

"Oh yeah," Jason said.

"Bennie was so embarrassed," Walter said, chuckling. "I remember when they displayed her picture as part of the exhibit as Female Astronaut of the Year. It was an exciting moment with all the distinguished guests and NASA alumni. She prided herself on being in control and orderly despite all the fanfare. But just as the picture was displayed, the security guard had to stop the show to rescue you four kids from the scaffolding you had climbed behind the display."

"I remember that. I convinced the kids it was okay to go up there to touch their mommy's picture."

Walter continued to chuckle. "Well, Bennie could not hide her face from the embarrassment. People stared at her, wondering how the Astronaut of the Year could let her kids run wild. I don't know if your mom ever told you guys, but the guard in charge of the exhibit lost his job because of that incident, and he blamed Bennie and Larry. Larry told me the guard tracked Bennie down when she and your mom were finishing their astronaut training and harassed her several times after that. Regretfully, Larry was dead serious when he threatened to use a biochemical weapon on the guard. Not surprisingly, though, the harassment came to a stop. To this day, Bennie is embarrassed to have her picture taken in public."

"Dad, come on."

"No, really. In fact, she visited Mom a few days before Mom's trip to the space station four months ago. Bennie was in the mission control room with Mom, but she refused to join in any of the press pictures with her. And when they took the prelaunch press photos of Larry and me last week, Bennie was there, but you won't find her in any of the pictures.

Space Reunion

Bennie is so sweet and supportive, but she is far from being a celebrity."

"Dad, that's all in the past. You and Mom are lifelong astronauts, period!"

"Well, kids, we will just have to wait and see what happens with this policy. I am sure Larry will bring the subject up to your mom in a few days when we dock with her on the space station."

"Is Larry also disgusted with this new policy?"

"Not really. Larry is too excited about his special mission to let any policy bring him down. He is looking forward to conducting the space-walk experiment and getting to ride on the space motorcycle. He can't wait to see the earth and outer space from such a unique vantage point."

"I'm sure you guys will have a blast playing around up there, but I still think this whole policy thing is a stupid idea," Jason said. "Well, I'm not going to worry about it. NASA has a few bureaucratic goons, but they will come to their senses soon. Anyway, Dad, I have a quick orbital mechanics question from class today. Can we discuss it before NASA cuts off our call?"

"Sure, son. Shoot."

"Okay, I'm out of here," Britney said. "I'll leave you guys to talk that boring technical stuff. Anyway, I don't understand why NASA wants to keep fooling around in space and screwing up the environment with all its experiments. Tell Mom hello for me and that I love her. Tell her I'm going to deal with The Metal Rocks tomorrow. She'll understand."

Jason started to hit the mute button again—he'd never seen Britney as such a loose cannon—but he decided against it.

"What was that all about?" Walter asked.

"I don't know. Girl talk, I guess," Jason said.

"Sounds fishy to me," Walter said. "See if you can get your sister back so I can talk to her before they cut us off."

"It's too late, Dad. She just dashed outside."

8

Earthquake and Strange Sky Markings

The next morning, the abrasive tone of his alarm clock forced Jason awake. He was eager to witness his father's launch and to learn more about the earthquake epicenter and if the sky scrolls were still there.

Redondo Beach is a beautiful town this time of year. The sun was just peeking over the neighbor's house through the mini blinds of his bedroom window in his parents' five-bedroom home. Walter and Jason had equipped their home with high-tech special features: cameras, burglar alarms, gas, fire and smoke detectors, and automatic shutoff systems for gas, water and electricity. In fact, they had high-tech gadgets dispersed all throughout their home.

Jason catapulted out of bed and dashed downstairs to activate the TV system in their converted TV/studio room. Since the launch accident back in 2003, interest in the launches was on the rise. He signaled Britney via the intercom to come down and watch their father's launch. She was still sound asleep after a late-night phone conversation with one of the band members and so was not interested. She had watched so many successful launches that they hardly carried significance to her.

Jason opened his instant oatmeal, added some water, and popped it into the microwave. He then grabbed a cup of peaches and a container of cottage cheese.

Jason spoke over the intercom, "Britney, do you want me to heat up your breakfast so it will be ready when you come down to watch the launch?"

No answer. Finally, after several attempts, she replied, "Yes. I will be down in a few minutes."

"Hurry down. We might see some of the special NASA video shots that might zero in on the astronauts and show Dad strapped into his launch seat. It also might show his strained, contorted face during the launch. It might be worse than Mom's during her launch four months ago."

As Britney finally made her way downstairs, she passed by the wide assortment of space memorabilia, achievements, and special recognition awards that were scattered along the wall of the staircase. She had always taken them for granted. There were pictures of her parents with dignitaries, including the president of the United States, and a cherished portrait of both her parents suited up with her sitting on their mother's lap and Jason sitting on their dad's.

"Hurry up, Britney!" Jason shouted from downstairs. He was already sitting in the media room that included a large +photon plasma TV connected directly to NASA's closed-circuit system. Their family had spent many hours together in this room sharing their televised experiences.

Britney went into the kitchen and grabbed the ham and cheese omelet, hash browns, and biscuits Jason had prepared for her. She began humming and nodding to a tune in her head.

The launch countdown had begun, and everything was proceeding without a hitch. When the countdown reached three . . . two . . . one, an enormous, brilliant plume of flame flashed across the wide TV screen, and the Falcon 15 lifted off the

Earthquake and Strange Sky Markings

ground in a ferocious display of power. The onboard cameras showed a quick pan of the astronaut's faces. Jason was delighted to see the contortions of his father's face and saved it on tape so he could tease him with it later. After the Falcon 15 had completed the liftoff phase and solid-rocket separation, it began the long coast phase before entering into the insertion orbit.

"It looks like things are going well and comfortable for this launch. I think we can safely head off to school," Jason said. He turned off the TV, and they left for school.

Everything *was* going fine as far as the public was concerned. But NASA was monitoring a slightly irregular heartbeat in Walter that had been detected earlier but felt it was likely an electrical artifact since it had not reoccurred. Nonetheless, NASA decided to keep monitoring it. NASA spared no expense in monitoring the vital signs of each astronaut for any potential problem.

Jason drove his SUV to UCLA while some of Britney's friends picked her up to head to CSULB. Two of the band members who were pressuring Britney to join their group were in the same ride pool. They were ecstatic over the news that promoters had selected them to go on a national tour with a major production company after the semester was over. For this reason, they put even more pressure on Britney to join them.

Jason entered the classroom at UCLA along with the rest of his classmates with the eerie sky scrolls still hovering overhead. They talked about the previous day's earthquake and news reports that indicated it registered only 4.1 on the Richter scale and that scientists were having difficulty finding the epicenter.

"Did you hear about the two GPS satellites that just vanished?" Austin said.

"Yeah! What happened?"

"They don't know, bro," Austin said. "Something weird is happening or has happened."

Brutus was still lying outside their classroom with yellow caution tape as a barrier. The ominous sky scrolls remained overhead, mesmerizing the students and faculty. An unquantifiable tension filled the air.

"It's obvious what they are," one student said. "The military has screwed up another one of their classified rocket experiments."

"I don't know about that speculation, but my father's launch got off okay this morning," Jason said.

Others continued speculating while waiting for Professor Thatcher to arrive.

**Despite Professor Thatcher's condescending comments about black people the day before, the competition for good grades at the university remained high. The recognition of academic excellence by one's peers was vital, and such competition often led to intense arguments and even fights. Graduate students believed they were the best of the best and had the egos to match. Clearly, they excelled academically and possessed remarkable determination, seeking to acquire the edge over other students by any means necessary. Many fraternity students managed to gain an edge by obtaining old exams from former students who had taken the same class, and some students worked in study groups that often sucked the life out of these hand-me-down exams. Historically, most of the professors had endured the same intense competition during their own student days. Consequently, they were less than

Earthquake and Strange Sky Markings

sympathetic about students suffering through rigorous and demanding hard work. In fact, many professors gained sadistic gratification from observing their students struggling and competing among themselves. They were proud to help foster intense competition, creating a savage atmosphere rife with conflict.

One student questioned Professor Thatcher's credentials. "You know, I have checked out the professor's record on the Internet and personal contacts and found that Thatcher's appointment at Harvard had come about through political means. Later, Thatcher was 'cast out' to UCLA because of problems he had instigated." Just as he was about to elaborate, Professor Thatcher walked into the classroom.

The professor had an Ace wrap protecting his left hand. He put his old-fashioned snap lock briefcase on top of his desk and opened it. He was still embarrassed and angry about his cowardly behavior the day before. With a stern look, he took out his neatly packed notes, books, and pencils and placed them on the lectern. Then he carefully removed a neat stack of forms and placed them on the desk.

"Class is now in session," he announced, peering through the thick lenses of his black-rimmed glasses. He fondled his curly sideburns.

The students quickly stopped chitchatting and noted a glistening light reflecting from his curly, white hair. He clasped both hands together, allowed them to drop to his sides, and then swung both arms up with his fists clenched. The students immediately gave him their undivided attention. He told them he had done some soul searching in regard to the concerns voiced by the students yesterday.

"Although I have taught school at Oxford and Harvard for many years, I need to be open to change," he declared. "Some great students have emerged through taking my classes and participating on my research teams. Even though most successful students have indicated my material is informative, thought-provoking, and even difficult, there is always room for improvement. All endured, and many have gone on to do great things. You made some valid points yesterday about the pace of the lecture. Accordingly, I have made a decision. *I am not going to change my lecture format one bit.* You UCLA students need to grow up and strive for the high standards maintained at places like Harvard, and you need to show respect for the legacy of my past students. Consequently, I have some dropout forms here for any student who can't handle the course load." He stared directly at Jason. "Does anyone want to drop the class without penalty?"

Stunned, no one budged.

"Well, then, let's continue with the lecture—*as is.*"

Professor Thatcher lectured at his usual sizzling pace, if not even faster. There were no questions asked during the lecture, and the class ended without incident. Afterward, Jason, Austin, and several other students stood outside the classroom, staring at each other.

"This guy must be crazy," Austin grunted. "Does he think he can get away with this?"

They strolled the campus grounds, viewing the towering trees and brownstone buildings, while continuing to berate the professor and trying to figure out what they should do. They observed several students staring uneasily at the peculiar sky scrolls when suddenly overhead there came a brilliant flash of light.

"Dude, did you see that?" Austin blurted out.

Earthquake and Strange Sky Markings

"Yeah, what was that?"

"I don't know, but it looked like lightning. But the sky is clear. No thunder clouds, nothing, except those damned sky scrolls."

"This is strange," Jason said.

"What do you make of it?"

"I don't have a clue. Lightning in California from out of nowhere?"

As Jason and Austin continued on to their next class, they noticed four students with shaved heads, wearing robes and sandals, standing on the grassy area and chanting, "These are the signs, these are the signs, these are the signs!"

More and more students began peering up at the sky scrolls. The mood of the mingling crowd moved from curiosity to uneasiness, concern, and even fear.

Later, as Jason and Austin drove home, they heard a radio report about the scrolls in the sky and the sonic boom.

"We have breaking news," the reporter said, interrupting their conversation. "The sky scrolls have been seen from Los Angeles to as far away as Houston, Texas. The Federal Aviation Agency indicated the scrolls were seen by several passenger jet pilots. However, they were above the altitude of the jets, which rules out the scrolls being any form of skywriting . . .

"In other news," she continued as Jason and Austin listened intently, "readings indicate the earthquake experienced yesterday at 11:30 a.m. measured 4.1 on the Richter scale. Many people felt it was much stronger than this and questioned the accuracy of this magnitude reading. Scientists have not yet determined the epicenter, but reports indicate they detected it over a wide area. Many people also reported hearing a sonic boom about the same time the earthquake occurred. This also was heard over a wide area. In other news, communication with two GPS satellites has been interrupted. In fact, the satellites are not shown on the radar screen or by telescope. NASA will be giving a briefing at 3:00 p.m. today.

"We now turn to the weather," another news reporter stated. "Several regions are reporting unusual sporadic lightning. Further, small sporadic tornadoes have been reported in two Southern California cities."

"Say, man, things really seem to be getting churned up," Austin said. "What the hell is going on? Somebody must have really pissed off Mother Nature."

"Like I keep telling you, Austin. This goes well beyond Mother Nature."

9

Girlfriend Apartment Visit

Later that evening, Jason headed over to Jane's apartment. While listening to the news, he was surprised to see what appeared to be several lightning strikes in the distance. He noticed the worried faces of many drivers and felt uneasy about all the environmental anomalies.

He was planning on helping Jane get ready for her trip to the university in Oaxaca, Mexico, to study the religious history of the Mayan culture. Her apartment was located within the densely packed apartment buildings throughout a twenty-block neighborhood adjacent to the campus that consisted of full-time college students. Cars lined the two-way, narrow streets on both sides. Parking was scarce and limited, so most students used underground parking or whatever street parking they could find. There was little green space to speak of, but all the buildings had a narrow row of tall, leafy spruce trees lining the front sidewalks. The earthquake had caused only minor damage in the area.

Jane shared a fourth-floor, two-bedroom apartment with her roommate Manuela, a stout, attractive Hispanic student who always wore a headband. She was polished, outgoing, tough and argumentative. Manuela advocated women's rights and was an officer in several organizations. Many believed she was on track to become a national leader. Manuela was also good friends with Britney. They knew each other from high school, where they both showcased their singing voices in the chorus.

Jane majored in cultural anthropology, and Manuela majored in sociology. They each had a small bedroom, but Jane's room had a small balcony where, through the trees, she could see a portion of the campus just over the top of the neighboring apartment building. She had obtained an extra security card with a password for Jason, which he used sparingly and never when Manuela was alone in the apartment. Manuela was slightly enamored with Jason but made sure she was never in a situation in which her feelings for him and her strong friendship with Jane might be put to the test.

When Jason arrived, Jane was still upset with Austin for convincing her to leave the cafeteria without finding out how badly the girl was injured who had slipped on the water that Jane had flung on the floor.

"Stop worrying," Jason said. "Why don't you call over to the cafeteria and see what they can tell you?"

Jane agreed and called the school. When she got through to personnel, she was told one of their employees had cleaned up the mess on the floor. They reported that the girl was not seriously injured, and as an apology, they had given her a free meal pass for an entire week. Jane was relieved but still annoyed over leaving the scene of a potentially serious accident.

"When I get back from my trip, I'm going to repay the cafeteria for covering up my misdeed," Jane said.

"If that would make you feel better, maybe you should do just that."

As they continued packing, Jason expressed his concern regarding Jane spending an entire month

Epicenter Uncertainty and Intimacy

with Luther; a classmate of Jane's who was also majoring in cultural anthropology. He was a member of the UCLA soccer team and a part-time model. One of his goals was to make it big in the modeling industry someday. Since Luther successfully flirted with many of the girls on campus Jason was certain Jane would be no exception if the opportunity arose.

Jane had been in several classes with Luther, and once before meeting Jason, in a moment of weakness, she had slipped a suggestive note to him. He was too busy being a playboy to respond to her immediately, but by the time he did, Jane had learned of his reputation with many different women and had lost interest in him. Clearly, Luther's good looks made certain things come easy to him. When Jane found out how irresponsible he was, she was really turned off. But Luther was still under the impression that Jane was still his for the taking. However, she now nearly despised Luther and was ashamed of her brief infatuation with him. She had opted to hide this entire encounter with Luther from Jason, including her note.

"What time is the bus leaving for the airport?" Jason asked. "I plan to come over early to see you off."

"Five thirty in the morning."

"Good grief, Jane! Leaving at 5:30 a.m.? That means you have to get up at 4:30 at least. Don't you think that's ridiculous?"

Jane stared at Jason without saying a word.

"Okay, okay, okay. I'll see you in the morning around five o'clock. I'll take you to the bus but let me finish helping you pack."

10

Epicenter Uncertainty and Intimacy

They continued packing her bags and watching the news, where reports indicated that seismologists were still having difficulty determining the epicenter of the earthquake. Further, serious communication problems were being reported with cell phone and Internet lines that had surfaced at the time of the earthquake. Reports were coming in claiming the earthquake had been felt from San Diego to as far away as San Jose.

"That doesn't make sense," Jason said.

"What doesn't make sense?"

"It's physics," Jason said. "The distance between those two cities is over eight hundred miles. Any earthquake felt over that large of a distance would have been massive at the epicenter, but there are no reports of any massive destruction or deaths." He paused again, thinking. "Although, you know," he murmured as a thought dawned on him, "it may be possible that there were two different earthquakes occurring about the same time."

The news reports continued, indicating that preliminary data showed the earthquake measured 4.1 on the Richter scale.

"That's bull!" Jason bellowed. "It had to have been more than 4.1! It shook the holy hell out of our classroom, and it was felt as far away as San Diego? Only a 4.1? No way!"

"In my classroom, we thought the walls were going to collapse," Jane said.

Epicenter Uncertainty and Intimacy

"For an instant, I thought this was the Big One people have been talking about for years."

"Me, too. I always worry that a lot of people will be killed by an earthquake."

"Don't worry about an actual earthquake because earthquakes don't kill people. Collapsing building do."

"In other news," the news report continued, "NASA continues the investigation of the mysterious disappearance of two GPS satellites. The Tavula volcano near the Solomon Islands has erupted, sending a plum of ash and debris thousands of feet into the air. Hurling rocks of fire and debris have pelted the area, trapping hundreds of local residents. These volcanoes, part of a chain of volcanic islands positioned in Micronesia, have been dormant for hundreds of years, and scientists are amazed and puzzled by its sudden eruption. The lava gushing out of these volcanoes created a spectacular display. One volcanologist indicated it was the most extensive lava flow he had ever seen. Rescue ships and aircraft are being summoned to the area."

The TV showed brilliant yellow-and-red jets of molten material shooting into the air. For miles around, smoke flowed high into the atmosphere.

Jason and Jane finished packing and began to relax, discussing the day's events. They had met two years earlier during Jane's junior year when she was a member of the UCLA intercollegiate debating team. Jason and Austin were walking by Ackerman Union Hall when they noticed several girls entering the building wearing T-shirts labeled "UCLA Debate Team." Jason and Austin followed the girls into the building, sat in the back of the auditorium, and watched the debate preparation. Jason watched Jane interact with her homely teammates, encouraging team spirit and creating a nurturing atmosphere that Jason admired.

"Austin, look at that girl, the second from the end. She is so beautiful. I need to meet her. She's in a class all her own. The other girls are too unrefined, boisterous, and disheveled for me. But she looks composed and sophisticated and gleams with confidence. Her shoulders and neck are proud, her blue eyes sparkle, and her luscious lips and face are so gorgeous and sweet that none could behold her without feeling she must be a descendent of an angel."

"Good grief, bro," Austin said. "Are you now a poet? She must be sent from heaven."

"Yep, she bears a cosmic stamp of approval in every one of her features. Is that the way you see it, buddy?"

Austin thought she was beautiful, too, but would not dare compete with his best friend over a woman."

"You're really focused on that chick," Austin said. "I'm going to trot over to the refurbished student center, play some video games, and seek out some fine, stacked chick like her for myself."

After Austin left, Jason remained behind, concocting a scheme to meet Jane. He noticed that she and her teammates fell short when it came to questions regarding astronomy and related scientific topics, so he decided to "accidentally" bump into her and her teammates at the end of the practice session. He would point out their academic weakness, divulge that this was his area of expertise, and disclose that he had developed a learning technique with high school students that had proven successful.

Epicenter Uncertainty and Intimacy

At first, everything went well with his strategy. He bumped into the girls and arranged for them to take his special tutoring sessions. Everything went according to plan except for one thing: all the teammates showed up except Jane! To make matters worse, the most "beauty impaired" teammate developed a heavy crush on Jason. Nevertheless, Jason endured and conducted two sessions of his special astronomy tutoring. Eventually, Jane showed up for the third and final class. Soon thereafter, Jason arranged to have lunch with Jane to make up the tutoring sessions she had missed. Thus, Jason and Jane's relationship began.

Since that lunch, their relationship had grown and they were together every day for almost two years. They had reached 552 consecutive days which they called JJ Days before Jane was slated to go on her trip to Oaxaca.

Earlier in the evening, Jane asked Jason to stop by one of the grocery stores in Westwood Village to pick up some seasoning sauce and wine for dinner at her apartment. When he arrived, he placed the sauce on her dining table while Jane sat up some plates on the folding trays in front of the TV where they frequently ate together.

Jane was an excellent cook of soul food, which she had learned from the maid of her grandmother. The aroma of smothered pork chops, cornbread, red beans and rice, seasoned collard greens, and yams made Jason's mouth water. Manuela often dined with them, but she was gone for two days to a school-sponsored conference on leadership and women's rights.

Jane complemented the soulful dinner with Jason's wine. As they started to dig in, they flipped from the game shows of *Jeopardy* and *Wheel of*

Fortune, which they usually watched, to check out the news.

Jason turned to the world news report, where the reporter stated, "Docking with the International Space Station is proceeding . . . In other news, earthquakes were felt in Arizona and New Mexico. We have received several reports of sonic booms and scattered lightning. Scientists are having difficulty explaining why there were no clouds in the vicinity associated with the lightning or tornadoes. The only cloud-like condition was the sky scrolls."

Puzzled, Jason stared at Jane. The news reporter continued.

"Furthermore, NASA reported the sonic boom was not caused by any known reentry craft. In other news, scientists still have not located the epicenter of the 4.1 earthquake in the Los Angeles area. The earthquakes reported in Arizona and New Mexico occurred simultaneously. Scientists claim they could not be the same earthquake because they are on different fault lines. They believe it was either a rare coincidence or that one of the earthquakes somehow triggered the other."

"Did you hear that?" Jason said. "Something is bizarre. To have that many unrelated earthquakes occurring at the same time is unheard of."

"What does that all mean—earthquakes, sonic booms, and lightning?"

"I don't know, Jane."

"I'm worried. Could we have screwed up the planet so badly?" she said, fear filling her blue eyes.

"Hey, don't get upset." Jason scooted over and put his arms around her. "Everything will be okay. I'm sure there is a logical explanation for all of this. I will be around for you no matter what happens."

Epicenter Uncertainty and Intimacy

Jason flipped the TV to smooth jazz and leaned over and kissed her gently on the cheek. Their eyes met deeply and he kissed her lips. Tenderly, he grasped her ponytail as she grasped his. Their arms gave way to full embrace. Their gentle nibbles escalated to passionate kisses and Jason's hand unconsciously slid under Jane's sweater. When he came to his senses and realized he was fidgeting with her bra, he slowly began to withdraw his hand. Jane gently grasped his forearm and softly urged his hand back. As Jason started to say something, she placed her finger over his lips and stroked his neck.

As desire grew, restraint faded. Jason unclasped her bra and slowly pulled her to him with both hands while kissing her gently on the neck. He gently lifted her sweater and pulled it off. His lips moved down her neck and into her cleavage. As her bra slipped down, his lips moved to her breast and gently cupped her nipple. She began undulating, her head thrown back as she moaned softly. The months of waiting were over. They looked deep into each other's eyes, knowing this moment would bond them forever. The sheer love and passion of the moment was theirs. Jane's bra slowly slipped to the floor as did her silk undergarments.

The background music was intoxicating as she reached and turned down the lights. The room glowed with love and desire. Her body called for him and Jason gave his love deeply to her as he took her. Now her heart was totally his while her mind spun in disarray from all he had given to her. They lay there quietly in a passionate embrace, sharing the warmth of each other's body.

"I love you, Jane," he whispered.

"Not nearly as much as I love you, Jason," she sighed, looking deeply into his eyes.

Aphelion

Jason spent the night filled with a bliss he could have never imagined and would never forget.

11

Pilot Boycott

Early the next morning, at the LAX airport lounge, six airline pilots were discussing the sky scrolls they had been observing for the past two days. Each of them was wearing a neat blue uniform with flying pins sparkling on the lapel. They were sitting on bar stools in their private lounge sipping coffee and waiting to depart for their respective flights to Chicago, Atlanta, and Saint Louis.

One pilot asked if any of the others were concerned about the sky scrolls. They all answered almost as one.

"Hell yes!" one junior pilot said. "I have to fly right under that stuff, and I'm concerned it may contain some toxic gas or biochemical agent or something."

"Oh, don't be so scared," a senior pilot said. "If it were something toxic, NASA or surely the FAA would have let us know right away."

"Yeah right!" another pilot blurted out. "They know what they're doing, just like they did during the Malaysia flight disasters. The FAA was in a shambles, man."

"Cool it," the senior pilot said. "With that attitude, you guys will send your passengers into a panic. Besides, the FAA would issue an alert at the drop of a hat if they had even a hint of concern. In fact, since 911 they tend to overreact to most things."

"I hope you're right," the junior pilot said. "These things are strange. Skywriting at altitudes above one or two thousand feet is unheard of. And all of our jets cruise at thirty to forty thousand feet. At that altitude, the gas or smoke usually used in skywriting instantly disperses, so we would see nothing. I don't get it. I think the government has some explaining to do. I just have to wonder if one of their experiments has gone bad."

"All these weird goings-on are a complete mystery to me," another pilot said. "I'm not sure if the scrolls are occurring where we fly. Are they in the eight-mile troposphere range, higher up in the fifty-mile stratosphere range, or in the two-hundred-mile ionosphere range where some of the satellites orbit? There might be some funky things taking place out there. I think we should organize our union to take action. Maybe we should boycott our flights until we have more confidence in the situation for our passengers, not to mention ourselves."

Everyone sat quietly, unsure of what to do.

"You guys can just sit here if you like, but I'm going down to the union headquarters and see what's what. Anyone care to join me?"

12

Departure to Oaxaca

The next morning, weighing on his mind was the possible end to his parents' space mission goals and Britney's apparent detour into rock music. In addition, he remained mystified about the unsettling environmental phenomena that had been occurring.

"Are you all packed now? I'll take your bags down to my car if you still want to go on this trip that should be canceled."

"Why would you say something like that?" Jane said. "I have to go for my class. Sure, I would rather stay here with you, but I have to do what I have to do."

They arrived on time and joined the rest of Jane's group in front of the bus. Jason was extremely apprehensive and nervous about Jane leaving now. With all the breaking news about the sky scrolls, the unexplained nature of the earthquake epicenter, and the other menacing environmental events, her leaving made no sense to him.

Some students near the bus were voicing nervous excitement about the upcoming trip. Others were standing nearby under the large eucalyptus trees analyzing the sky scrolls that had not dissipated overnight while others were threatened by the surprise lightning strikes. Jane began conversing with some of her friends about artifacts in Oaxaca and hinting about her blissful evening with Jason.

Jason walked over to the faculty guardian who was traveling with them and started a conversation. He suggested that the trip be canceled or at least rescheduled because of the present environmental safety issues. The guardian was also uneasy but was reluctant to consider canceling since she had worked so hard to plan and implement it. Overhearing the conversation, a neatly dressed, perfectly manicured, wavy-haired guy with a muscular physique approached and butted into the conversation. It was Luther.

"Relax, science boy," he said. "Why don't you back off, dude? I'll take care of your little girlfriend."

"Why don't you kiss my ass?" Jason snarled.

The insults escalated into a shouting match. Things really got out of hand when Luther put his hand on Jason's shoulder, and Jason slapped it away. They both then yelled louder, grabbing each other. Jane, the guardian, and the other students jumped in to break it up. Jane was embarrassed, and the guardian threatened to kick Luther off the trip. Everyone finally calmed down. Jane and Jason respectfully gave quick good-bye kisses as the students all boarded the bus.

Watching the bus depart, Jason nervously waved good-bye to Jane as a troublesome thought plagued him: Luther might try to apologize to Jane by putting me down and then will attempt to seduce her.

As his stomach churned from anxiety, Jason walked to the cafeteria to get some breakfast before class. When he arrived, he saw some of his buddies sitting around a lounge table drinking hot chocolate and eating breakfast. He walked over, plopped his backpack on the table, and slumped down into a chair. "What's up?" he said.

"Hey, did you hear about the astronauts onboard the space station?" Austin said. "They saw the sky scrolls, too, and confirmed that the bands exist up there."

"Yeah, I heard it on the news," said one atmospheric science student. "They said that NASA scientists believe the scrolls are some form of unexplained exoatmospheric particle collection caused by condensation."

"What the hell does that mean?" Austin asked.

"Look, stupid," the thermal science student also sitting at the table replied. "Environmental scientists from the EPA are being assembled to determine possible causes."

"Does the EPA even employ real scientists?" Austin said.

"Yes, dummy," the thermal science student replied. "Some of them say it may be from a solar flare or some other phenomena. They plan on obtaining some samples from the sky scrolls to examine their composition."

"Yeah, right!" another one of their buddies blurted out. "I think you guys are all screwed up. It's clear the scrolls were caused by aliens."

With that, everyone looked at each other with a mix of amusement and puzzlement.

13

Professor's Accusations

That afternoon, Professor Thatcher, after viewing the menacing overhead sky, scampered into the office suite of the chairman of the UCLA engineering department. He greeted the secretary.

"Did the earthquake on Monday shake you up at all?" Thatcher said.

"Yes, it scared me!" she said. "I went and stood in the doorway. We all waited awhile, and then we went back to work."

Her desk was positioned as if it were barring the entrance to the chairman's office.

"I need to see the chairman right away regarding an important disciplinary matter," Thatcher said.

"I'll let him know."

She called the chairman on the intercom, stating he had an unscheduled faculty member who requested a meeting right away, and a brief dialogue ensued. She then nodded her head and sent the professor straightaway into the office.

Chairman Jacob Chapman sat in a large reclining black leather chair, reading some confidential information about the upcoming retirement of the school's current dean of engineering. Chapman was scheming for this promotion.

Chairman Chapman tried his best to conceal his potbelly bulging over his belt-cinched pants. His receding crew cut gave him the appearance of a middle-aged military man. Articulate and quick-

Professor's Accusations

witted, he seldom conceded to any view that was not his own. If there ever was a conniver and manipulator, he fit the bill; consequently, he prided himself on delegating every task to others to work hard for his benefit.

"How are things going, Professor Thatcher?" Chairman Chapman said. He was under the mistaken belief that Thatcher had important connections with UCLA's top academicians since Harvard awarded him a visiting professorship sabbatical grant to teach there, as this was rare. He leaned back in his chair, stroked his crew cut, and peered over his gold reading glasses attached to a cord around his neck. "Have a seat. What happened to your wrist?"

"I have some bad news. I had a major problem in one of my classes the other day."

"Sorry to hear that. What happened?" the chairman began to frown while fiddling with the cord.

"A student, Jason Scott, he . . . well . . . he attacked me in class."

"What??"

"He attacked me, slammed my hand on the table, and caused me to plummet to the floor and then called me a pig."

"You mean Jason Tyler Scott, our research student? The one in charge of the Galactic System Project?"

"Yes!"

"I can't believe that."

"Well, it damn sure happened, and I hope you don't let the little bastard get away with it!"

"Calm down, Professor Thatcher. I am just stunned, is all."

"I have been working hard to expand the horizons of these students, but you UCLA folks continue to disrespect me."

"Professor Thatcher, I don't appreciate the tone of your comments about our students and my faculty."

"I am mainly referring to the students. I am tired of these bratty kids from big-time astronaut families pushing me around and getting credit for everything."

"Hold on a moment. I thought you held Jason's parents in high esteem with their role in the International Space Station?"

"Don't fool yourself. Both of his parents simply take their achievement for granted. My parents worked their butts off in England their entire lives hoping I would have the simple opportunity to meet an astronaut. Most of you UCLA folks don't know what hard work is."

"Hold it. Let's leave personal feelings out of it. We at least stick around long enough to complete what we start."

Professor Thatcher shot him a fierce look.

"Ignore that comment, Professor Thatcher. That was not fair. I know your credentials and track record are solid."

Thatcher wondered exactly how much Chairman Chapman knew about his track record.

"What precisely did Jason do?" the chairman asked as he stood up.

"Well, Mr. Chairman, apparently he is an impatient student who does not like my methodical lecturing style. He became enraged after I rebutted his classmate's attempt to redirect the entire lecture to something only they selfishly wanted to hear. He then began criticizing my lecture."

"Good grief! That doesn't sound like Jason. What would provoke that kind of behavior?"

Professor's Accusations

Thatcher sprang up and shook his fist at Chapman. "I don't have to take that behavior from a spoiled student. If you don't do something, I'll file an assault charge on him."

"Just hold it there. Let me take care of it. Why don't you calm down? Take a few minutes to file a report, and give it directly to me. I assure you I will take the appropriate disciplinary action."

In the adjoining office, Austin was preparing to use the fax machine and overheard the entire conversation. Having witnessed the so-called tumbling incident, he was stunned at Professor Thatcher's distorted account and outright lies. "What a chicken-shit bastard!" Austin mumbled to himself. "We should have hooked him the other night and now somebody has to kick his ass."

Rather than barge into the office and make a big scene, Austin tiptoed out of the office unnoticed. He was fuming at the possible consequences to Jason if the chairman were gullible enough to act on Thatcher's accusations. Austin was torn: should he tell Jason, refute the story to the chairman, challenge Professor Thatcher face-to-face or get his buddies to really accost him later? As he stewed over which course to take, he grew more and more angry and kicked over a trash can on his way to the library. He then glanced at the eerie sky overhead.

Back in Chapman's office, the chairman politely asked Professor Thatcher to return to class. As Thatcher closed the door behind him, Chairman Chapman thought what a mistake it had been to accept Professor Thatcher at UCLA as a visiting professor. Now the school was in jeopardy of loosing recognition for the student Nobel Prize nominee before they even finished processing the paperwork. If they withdrew Jason's nomination, the financial support they were receiving from the alumni foundation would also be at risk, and Chapman's chances of ever making dean of the engineering school would erode.

He opened his desk drawer and pulled out a pack of mint-flavored gum. He pivoted around in his chair, stood up, and reached for two books at eye level from his floor-to-ceiling, oak-and-mahogany bookshelves. He pulled out two physics books with one hand and reached deep behind them with the other. Checking to make sure no one had walked into the room, he cautiously pulled out a half-empty fifth of Jack Daniel's, removed the top, and took a gulp. Sighing, he thought, This job is going to kill me.

Chairman Chapman returned the whiskey to its hiding place and popped three sticks of gum into his mouth. He then sat back down in his big chair and contemplated what he might have to do about Jason.

14

Professor Assaulted

Later that day, Austin was still fuming over Thatcher's deception. Austin's dreams of being part of a Nobel Prize team with Jason were dashed. Would his family now forever think of him as a mere girl chaser?

As he was leaving the library, Austin spotted Professor Thatcher strolling to the gym. Austin still wasn't sure if he should confront the professor or go back to Chairman Chapman and let him know what really happened. He didn't know why someone would want to cause such harm to a good person like Jason, though he *sometimes* showed tendencies towards arrogance.

Austin followed the professor into the gym. There appeared to be no one else in the locker room, so Austin concealed himself behind a row of lockers. Gnashing his teeth, he considered the severe damage the professor had just inflicted upon the career of his best friend. With his fist clenched, he moved from behind the lockers to confront the professor.

Thatcher opened his locker to get his gym clothes. Austin stood motionless, watching as the professor began to get dressed. As he bent over to tie his shoes, his rear end protruded toward Austin. In a fit of rage, Austin ferociously kicked the professor's rear. His behavior startled even himself, and he instantly ducked back behind the lockers.

The force of the kick caused the professor to lose his balance, and he toppled over the bench. Stunned, he tried to protect himself by covering his head with both hands before hitting the floor. As he tumbled down, his thick glasses flew off his head and slid across the floor. He rolled over on the floor, confused, stunned, hurt, and began emitting his girlish squeal.

Austin's heart pounded as he thought, what the hell did I just do? He listened to the scuffling of the traumatized professor and managed to stay concealed behind the lockers before escaping through the side exit of the locker room. When he made his getaway, he felt both elated and nervous.

Austin frolicked down the hill toward the cafeteria and arrived with his heart still pounding. He swaggered into the cafeteria with the sounds of Thatcher's groans and squeal still echoing in his head. Austin's bewildered buddies watched as their giddy friend danced around their table like a kid, his head bobbing from side to side, giving high-fives to his befuddled buddies.

Suddenly, he looked across the room. Lumbering toward him was a silhouette resembling Professor Thatcher.

15

Space Station Redirection

Walter and Larry were on their way to rendezvous and dock with Dorothy. I'm not sure if I should be candid with Dorothy about Jason's disrespect in school and Britney's strange comments about the rock group, Walter thought to himself.

Larry started communication with the space station. "We are now in orbit and expect rendezvous at fourteen hundred hours. We are looking forward to the docking process and joining Dorothy and the other crew members tomorrow."

Larry was looking forward to the rendezvous since he was scheduled to make an historic space motorcycle ride. Inspired by Larry's childlike excitement, Dorothy had sketched a caricature of him riding the space motorcycle as though he were strapped to a bucking bronco and having a great time. She had also talked to Bennie about his excitement and, upon their return to Earth, promised to give her the sketch of Larry as a flamboyant space cowboy.

NASA had communicated to the space station the news and speculation concerning the sky scrolls. As Dorothy traversed through the weightless confines of the station, she peered out the side viewing port into deep space.

"Hey, guys. I am looking toward the bottom right in the viewing area. I see the edge of something that should not be there. At first, I thought it was part of the aft solar arrays connected to the center tube structure of the station. I know the arrays span the length of an entire football field, but this is not they. Let me maneuver to a different viewing port to get a better look at whatever it is."

Dorothy held onto one of the handrails that traversed the interior walls of the station to aid in her weightless maneuvering through the diameter of the tubular walkway structure. The interior was about the size of a passenger cabin of a 747. The next portal window afforded a better view.

In a slightly panicked voice, Dorothy said, "I see a thin beige layer of matter shaped like a blanket with a matrix of dark lines." As she moved closer to the port, it suddenly dawned on her: These are the sky scrolls everyone is talking about! There they were, right before her eyes.

Dorothy shouted to the others, "Look out the starboard portal over North America. This must be the infamous sky scrolls up close. They resemble the aurora borealis, except they seem to pulsate with life and to be encircling a large sector of the earth."

Another crew member peered out another viewing port. "I see them. There appears to be several thin, faint, scorched, transparent-colored bands that seem to contain some embedded scribbling."

NASA officials had no credible explanation for the phenomena. For that matter, neither did the astronauts. No clouds existed so high up, yet there they were. As did the airline pilots, some of the astronauts expressed major concern about the scrolls.

On Earth, more and more reports were streaming into the news media about the sky scrolls, the sonic boom, and the increased lightning and tornado events. This inexplicable situation baffled NASA, prompting it to impose a news blackout to avoid possible worldwide panic. The fears expressed by some of the astronauts leaked to Canadian news sources and quickly spread to all the international news stations. NASA scientists were encouraged to downplay the phenomena by explaining them as aurora borealis. The scientific segment of the public was speculating that the sky scrolls were some comet tail material while the political segment claimed an international terrorist plot.

Some amateur astronomers in Sedona claimed they could make out the scribbling within the sky scrolls. They claimed the scrolls were composed of numerous writings, and they could clearly see the words *science* and *God* within what they claim was actual text. Closer examination revealed what appeared to be words. They claimed that anyone with a good pair of binoculars could see words inscribed on the scrolls. This revelation sent a shock wave throughout every community.

As the news and speculations continued to spread, people from all over the world began pointing their binoculars and telescopes toward the scrolls. Local reports continued to pour in that actual words appeared within the scrolls. This news quickly spread. Leaders in China claimed they could make out some Chinese words within the scrolls, *science* and *God.* This set off a wave of turmoil throughout the religious communities around the world. Signs and messages began appearing on the streets, claiming from the book of Revelation, "Jesus is coming!"

16

NASA Chief Intervenes

Dr. Norman Braxton, the head of NASA working out of the mission control center in Houston, received feedback from the astronauts on Wednesday and initiated quick proactive measures at a news conference that evening. The Falcon 15 was preparing for docking with the ISS while things on the ground were unstable. He announced the assembly of a team of top scientists to investigate the phenomena.

Dr. Braxton promised to keep the public apprised of the latest developments. He cut a dashing figure in his three-piece suit and was eager to take on a new challenge. He was ambitious. His goal was to be recognized as one of the most renowned leaders ever in NASA. Nevertheless, several people on the president's staff and many religious leaders did not trust information coming from Dr. Braxton. They considered him an arrogant, politically astute technical administrator who was always scheming to associate with important people.

During his rise in the ranks of NASA, he received much unwanted attention for an article he published in a well-known scientific journal. In the article, titled "Scientific Explanations of Miracles," he argued that all alleged miracles ultimately had a scientific explanation. His deduction was that all phenomena must ultimately have a rational basis. This premise essentially dismantled the notion of sainthood since Catholicism dictates that at least three miracles be ascribed to saintly candidates.

Further, many claimed Dr. Braxton rejected the value of prayer and even the existence of a Supreme Being. Although he had a strong religious upbringing, his faith drifted during his intense technical training at Yale University. He stubbornly clung to his nonreligious position, which kept him in the limelight and paradoxically contributed to his rapid rise in NASA. But his divisive public comments created a growing backlash within the religious community, the depth of which he was oblivious.

His key technical staff members were assembled in the mission control room. Once everyone had arrived, Dr. Braxton surveyed the bank of video displays positioned on the walls. Several TV screens hung from the ceiling. One bank of TVs displayed all the space station cameras capturing the activities of the astronauts while other monitors showed the activities of key launch facilities throughout the country.

All seven members of his staff sat around the elliptical, mahogany conference table.

"Dr. Zelman," Dr. Braxton said, addressing the highly acclaimed director of astronomy. "What do you think is happening with this sky-scroll business, and why did two GPS satellites simply vanish?"

"To be honest, it beats the hell out of me."

"Well, that's not good enough!" Braxton leaned forward, placing both hands on the table.

"What do you mean, that's not good enough?" Zelman fired back.

"We will look like a bunch of idiots if we don't do better than that."

"Oh, so it's my fault?"

"No. Our entire organization has lost credibility since the Space Shuttle disaster."

"So you're saying it's NASA's fault?"

"Dr. Zelman, don't be so damned defensive. I am not blaming any one person or department. It's just the public and the president have had little confidence in any of us since that disaster. Our own astronauts are publicly rebelling against our new policy limiting husband-and-wife teams in space missions. We are going to look like a bunch of bungling idiots if we don't get a handle on this."

"Sir, maybe you should stop worrying so much about image and concentrate on figuring out what is going on. People are terrified."

"What do you suggest?"

"Why don't you do something like redirect the planned assignments of the Space Station crew to examine the scrolls rather than allowing them to continue their regular mission? Maybe they can help determine whether there is actually writing within the scrolls."

"Oh, don't give me that crap about writings in the scrolls," Dr. Braxton replied, grimacing at the absurd notion.

"Well, Dr. Braxton, the facts are there."

"Dr. Zelman, sir, the fate of our nation is in man's hands and no one else's. It is up to us to figure things out. We are scientists. We need not rely on voodoo or anything of the sort."

"Sir, with all due respect, just because you believe science and technology alone can answer all questions doesn't mean the rest of us feel that way. Our Maker may have some small part in determining the course of our lives."

"You're free to harbor your personal beliefs, but let's not get too biblical here."

"Biblical or not, you need to keep an open mind, Dr. Braxton."

"Enough of that. Let's deal with what we can. I'll take your advice and redirect the crew to begin examining the scrolls. Are there any other great ideas you brilliant folks think might help?"

There was silence, as everyone sensed it would be pointless to introduce any other constructive ideas in the face of such skepticism.

After performing the proper NASA protocol, Dr. Braxton sent out a directive indicating that due to all the concerns regarding the composition of the sky scrolls, reassigning Dorothy and her crew members was necessary. They were to suspend ongoing Space Station experiments and begin making special observations. Then they were to start taking measurements of the sky scrolls as soon as possible after the upcoming Falcon 15 docking. With any luck, this would help determine the underlying cause of these mysterious occurrences.

Unbeknownst to Dr. Braxton, his assistant had already taken the liberty to instruct the crew to begin expending precious resources to focus on deciphering the words within the sky scrolls.

17

Valium Overdose

Back at UCLA, Jason was thinking that Jane should have arrived in Oaxaca, Mexico. It was Thursday, three days after the earthquake, and the persistent sky scrolls and the unexplained sporadic lightning were wreaking havoc throughout the country. He was concerned about the impact the anomalies might have on air travel and on Jane's flight in particular. Today was also the day his parents planned to dock in space.

He picked up the phone. "I want to find out if flight MA141 has landed in Oaxaca, Mexico. Can you provide me with some information?"

"Yes, we can, sir. Let me pull that information up on my computer."

Jason waited nervously.

"Sir, my records show that flight MA141 has landed safely in Oaxaca, Mexico, and all passengers have deplaned."

Relieved, Jason said, "Thank you very much."

The faculty guardian helped escort the students to their dormitory. Jane was assigned to a room on the second floor with two other girls and three guys, including Luther. The students were well aware of the sky-scroll hysteria and heard rumors the unexplained earthquake was worldwide. They were told a rare and unusual tornado had damaged one of the planes on the far runway and they were blessed they landed okay.

Valium Overdose

Jane had not been able to get through to Jason since all the phone lines were jammed. She took some Valium to calm her nerves, forgetting that she had taken some earlier in the day after the scuffle between Jason and Luther. As her fellow students enjoyed a snack in the campus cafeteria, Jane became groggy and disoriented.

"Jane, you look exhausted," one of her classmates said. "Are you okay?"

"Yes, I'm fine. I'm just a little tired and anxious from thinking about all the things that are happening."

Luther, noticing Jane's odd behavior, said, "I agree, Jane. You look a little woozy."

"I'll be okay," she said.

"You really ought to take a break for now and head back to your room," one of her friends said.

"I'll be happy to escort you back to the dorm," Luther offered. Some of Jane's friends offered to accompany them. "No, don't worry about it," Luther insisted. "I can manage. I need to go back to the dorm anyway."

Luther and Jane spent the rest of the afternoon in their dorm suite together.

18

Awesome Universe

All day Thursday, Jason immersed himself in his research work, in part to avoid worrying about Jane and also to complete his studies. Along with Austin, two other research students, and some lab personnel, he collected data to support their model of the movement of the Milky Way, which was part of the Galactic System Project. Their models mapped the relative position of Earth's solar system with respect to the entire galaxy. They had formulated a theory based on the premise that the universe is expanding about a mystical fixed point in space and that all matter, including the galaxies, continues to move away from this point, perhaps someday to return.

Jason glanced at the breaking news feed and alerts appearing on his monitor and commented to his colleagues, "Reports are coming in from London, Lima, Sydney, and Tokyo indicating they each has experienced earthquakes—all registering about 4.1 on the Richter scale—at exactly the same time on Monday."

"It's clear to me that these developments indicate a worldwide earthquake phenomenon has occurred with no logical explanation," Austin replied. "The sky scrolls are widespread but are limited to southern Canada, the United States, Western Europe, and the northern portions of Africa."

"That's right," Jason said. "It says here that scientists from several independent groups are turning their focus to the links between the earthquakes, the sky scrolls, the sonic boom, and the sporadic lightning and tornadoes. Apparently, there are unexplained tornadoes and lightning strikes in nearly every major city in the northern hemisphere."

Austin joined in reading the news. "It seems that numerous communication lines are down or interrupted, making it difficult to impossible to verify and confirm the entire calamity that is taking place."

Another student commented, "The National Weather Bureau forecasts are indicating their satellites consistently need correction and updates because the orientations of their satellites are continuously shifting. We know the weather satellites are positioned in a fixed geosynchronous orbit around the earth, which means they stay over the same spot over the earth all the time. To ensure they stay there, the satellites look at certain stars to adjust or verify their position relative to positions on Earth. It appears their positions have shifted over the last few days. They are saying this was the case with all six of the satellites composing the weather satellite constellation system."

Another one of Jason's colleagues said, "The other satellite systems also have been requiring increased recalibrations, including Pan am Sat and those of the GPS system. These geosynchronous satellites also used the stars to point them in the right direction. They are completely dumbfounded as to why two of the GPS satellites have vanished."

The unexplained positioning-shift phenomena and missing satellites had the world's top scientists completely baffled and scrambling to find answers. They were speculating that there was some connection between the sky scrolls, the worldwide earthquakes, the satellite orientation problems, and the missing satellites. Some on the team had even considered the Gaia Theory in which the Earth is not just teeming with life; the Earth, in some sense, *is* life. This revolutionary idea claims everything happens for an end: the good of planet Earth.

Jason pointed out, "We are fortunate we are able to use our research model to predict the positions of celestial bodies accurately. Our focus has been on assessing complete star systems revolving within our galaxy; however, since our simulation can be applied to these local satellites orbiting the earth, we should do that for the time being. What do you guys think?"

During their discussion, Tyresha, the curvaceous lab secretary, overheard their conversation when she walked into the lab spouting "I don't know how you weird people understand all that technical stuff," she mumbled.

"It's not that difficult," Jason replied, looking at her.

"Sure, honey! It's as clear as mud," Tyresha stated as she sauntered past the guys.

"It's not that difficult, Tyresha. There are many simple analogies that explain what we are doing."

She planted both hands on her voluptuous hips. "Okay, honey. Let's hear one of your simple analogies."

"Okay, Miss Tyresha. It's like using an ATM machine to get cash."

"You mean an ATM machine to get money?"

"Yes, that's right. When you put your card in, it electronically reads information from the card. It then sends information about you to a central computer located at the main center. This computer searches your name and finds your records with your current balance. It adjusts your records and sends a message back to your local ATM that gives you your eighty dollars or whatever."

"Okay, honey, so what's the connection, Mr. Analogy Boy?"

"Well, with our celestial model, we input information about a particular celestial body. For example, a GPS satellite."

"What's a GPS?"

Jason pondered whether he was wasting his time. Then again, he was often accused of not taking the time to help less-enlightened students. "It stands for global positioning satellite—the thing they use to find a stolen car or give directions in cars."

"Honey, my car had better be in the parking lot."

"Seriously, Tyresha, the information we request, for example, may be this. Where is the satellite now? Or where is it going to be later at a given time? This is like requesting from the ATM how much cash you want and finding out how much you have left. Are you with me?"

"Keep it going, My Love."

"Okay, sister. It looks like I'm getting through," Jason said with a laugh. "In our model, this information goes to the computer memory file and looks up the name of the body of interest—say, the GPS satellite. Once the name is found—just like it finds your name at the ATM—it does some calculations. It then determines where it is located—just like your balance—and calculates where it will be located in the future—as with your balance after the withdrawal. This information is then sent back to us. That's all there is to it."

"That was an excellent explanation, Mr. Analogy Boy. I think I understand your model now. Everything is copacetic. Thanks again for taking the time to explain it to me."

"You're welcome," Jason said with a nod.

"By the way, there is one favor I want to ask. Can I use your model and input a request for sixty dollars and pick up the money here?"

The other students looked at Tyresha with puzzled frowns and shook their heads as if to say there was no use trying to explain it to her.

"Oh, never mind," Tyresha smirked and walked off.

"Hold it right there, Tyresha," Jason called after her. "Come back in here." Had she understood anything? "Come over here to our computer monitors, and sit your curvy butt down."

"I beg your pardon?" she replied.

"Oh, just sit down. You know what I mean. I am going to crank up our research models and let you see with your own eyes what we are talking about."

"What do you mean?"

"I'm going to put on a show for you. Austin, would you turn on all the monitors and boot up the computer workstations so we can execute the full research simulation? After that, go down the hall and get the other two department secretaries since we promised to give them a demonstration one day. I think now is the time."

"Yes, Bwana Tarzan. I'll get right to it," Austin replied. "Anything else, boss man?"

"C'mon, Austin. Get the lead out, would you? *Please*, sir."

"I'll be right back with your request, boss," Austin said, jumping up and saluting. "I'll be right back with the other secretaries, boss."

Jason and the others waited while Austin headed down the hall.

"Tyresha, please come over here and sit in front of the monitor for a moment. Now be quiet. Shush," Jason whispered as he surveyed everyone in the room. "I want you to listen for sounds. Forget about all the things going on outside for a moment."

Everyone in the room wondered what he was up to. Nonetheless, they cooperated, and the room grew quiet.

"Please look out the lab window where the sunlight is coming in," Jason said in a low voice.

Although puzzled, they did as he said. He asked them to tell him if they noticed any flickering of the sunlight or heard any rumbling. The room was quiet, motionless, and calm.

"Great!" Jason declared. "Did anyone notice how serene everything was?"

"Yes, everything was very quiet. But I did not notice anything unusual," Tyresha replied. "What's the point of all this, Mr. Analogy Boy?"

"Be patient," Jason said. He put his hand on his chin and nodded as if deep in thought. "What you folks perceive as reality is not reality," he whispered. "The reality is there are enormous cataclysmic events of unimaginable magnitude occurring at this instant, but you are not aware of them. When Austin comes back, I will show you what I mean."

Everyone sat quietly, thinking about Jason's words. The air was charged with suspense. Then Austin walked in with an entourage of support staff consisting of two secretaries, a female classroom custodian, and two cute female undergrad students who just happened to be in the nearby office.

"Good grief!" Jason said, annoyed by all the uninvited people. "Austin, you always have ulterior motives for every simple request."

"No, not really. I merely adjust to the situation. Demonstrations are an opportunity to meet cute girls. That makes perfect sense to me."

"Okay, no problem," Jason said. He surveyed the roomful of curious faces. "Everyone gather around here, and let's proceed."

Jason turned on the large computer monitor and selected the virtual space probe to move around the universe. The probe on the screen moved toward the sun initially seen from a distance as we would normally see it.

"Here is the sun, and it looks pretty benign from this distance," he said. "But look closely as the space probe zooms in on it. Pretend we are on a protected space capsule going straight into it, and you will notice some profound changes.

"First, we can see the sun's surface is incredibly active and violent. There are countless nuclear explosions going off every second, one on top of another. Even the smallest of these explosions would dwarf the force of all the nuclear explosions ever experienced on Earth combined and would annihilate our entire planet instantly. No living thing—animal, mineral, or substance—could survive for an instant on the surface of the sun, as unimaginable high temperatures and extreme energy bursts exist there. The sun is clearly not as calm as one might think when viewing it from a great distance. We can't hear the sounds because there is no air or atmosphere to transmit them, just radiation that we experience as heat. But let's move away from the sun in this simulation to examine some even more startling occurrences."

He clicked the icon labeled "Solar System."

"As you can see, our sun is the only star in our solar system. It would take our space probe several minutes traveling at the speed of light just to reach Pluto, which is no longer a planet but a gaseous body in our solar system. Specifically, traveling at 186,000 miles a second, it would take five-and-a-half hours for light from our sun to get to Pluto."

"I always consider light just there whenever I see it. I never think of it as having to travel to get to me. I guess it's like those big firework displays on the Fourth of July—you see the light before you hear the sound," Tyresha said.

"Well, sort of," Jason said. "When a firework goes off, there is a delay before you hear the sound. In reality, there are two delays—one from the light of the explosion and another from the sound of the explosion. Sound travels about seven hundred miles per hour while light travels at over 186,000 miles per second. Thus, we can detect the sound delay but not the light delay from a single explosion. Light can go around our entire earth three times in just one second."

"So you're saying if we were on Pluto, and there was a big explosion on the sun, it would take five hours before we would see it," one of the secretaries deduced.

"Yes! That is precisely what Einstein was talking about in his theory of relativity. Let's leave all that alone for now and get back in our space probe."

Everyone nodded. They were still with Jason so far.

"Look at the screen now. As we move farther away at hyper speed, peering into the night sky, we notice many other suns or stars. I mean many, many more. Our violent sun is a relatively small, calm star when compared with some of the other star suns. With the orbiting Hubble Telescope, along with several huge ground telescopes, we are now able to peer into deepest space." Jason moved the space probe to a dim spot on the computer screen. "Watch. I'm going to move in on this harmless little speck, enabling us to take a closer look."

An enlargement of the speck suddenly appeared.

"Look at it now. As we examine things more closely, we notice that small speck is actually another star—a *huge* star. In fact, this giant red star is so large that our entire solar system could fit inside of it."

"This is one big, bad dude!" Austin interjected.

"Good grief!" said one of the cute girls. "That's amazing!"

"Yes, honey, it's really big," Jason said. "We have captured actual images from the Hubble Telescope in our simulation using a combination of animation and mathematical computations that closely simulate what scientists know. How many stars do you think there are in our galaxy?" he asked, looking at Tyresha.

"Are you asking me? Hell, I don't know . . . thousands, maybe even a million?" she replied.

"Too low," Jason said. "Way too low."

"Darn. A hundred million stars?" one of the female students blurted out.

"Still too low," Jason said. "There are fifty to a hundred billion stars in our galaxy. Let me show you something that will blow your mind." He directed the space probe to a dark region on the screen and began moving in. Some dim specks began to come into view. "Look closely. These stars are not stars at all. Watch as we move in closer. This one speck of light is actually an entire galaxy that itself consists of billions of stars."

"Are you for real? That's amazing!" Tryesha muttered.

"Yes, this is for real. Now just hold on. I want to show you one last thing. When we look out into the night skies, and see the stars and other specks of light, some of which turn out to be galaxies, astronomers now believe there are many of these. Let me ask you something. How many galaxies—not stars but galaxies—do you think there are in our universe?" Jason looked at Tyresha again.

"Oh, come on, Jason. I don't have a clue. Maybe a thousand?" Tyresha said.

"No, many more than that. We are not able to see most of the other galaxies with the naked eye," Jason explained.

"Okay, so how many galaxies are there in the universe?" asked one of the female students.

"One hundred billion of these little specks are out there, each representing a different galaxy. And again, within each of these galaxies are fifty to one hundred billion stars. I don't know about you, but that is absolutely mind-boggling to me. Collectively, there are seventy sextillion stars in our universe."

Everyone sat stunned, beginning to grasp the awesomeness of the universe.

"Well, Mr. Analogy Boy, let me ask you a question," Tyresha said. "What is at the end of the universe?"

Jason, Austin, and the other research students looked at one another. None of them knew the answer.

"You ask an interesting question," Jason said. "Modern science says nothing is because the Big Bang theory indicates everything is expanding from a single point and nothing has gotten out there yet."

"I'm asking what *you* think, not some other scientist," Tyresha insisted.

"Hmm . . . I think our universe is just one of many. In fact, I think it is just a speck itself, and that there are hundreds, perhaps thousands or millions, of universes out there. Scientists estimate our universe to be almost fifteen billion years old. I believe there may be other universes much older, perhaps a hundred billion or a thousand billion years old, each located far, far away from ours."

"That is simply astounding, Jason. What do you call all of these universes?"

"You just said it: 'allniverse.' That's what we have been calling all of these universes. Our universe is simply a speck in the allniverse. That is why I always say 'our universe' and not 'the universe.'"

The other people in the room sat frozen with astonishment.

"Anyway, that's just my opinion. But let me end this. I think that may be enough complex astronomy for the time being. Let's get back to our research model. All of these stars, planets, solar systems, and galaxies are not just sitting there. They are moving, interacting. In fact, the Hubble took a picture in 1998 of two galaxies that are colliding as we speak. This means that the billions of stars within the two galaxies are smashing into each other."

"Awesome, baby, awesome," Austin said. "Can you imagine if another star smashed into or collided with our sun? What all would happen?"

"You're right, Austin," Jason said. "Now we have to let our audience know that's where we come in. We're collecting the images from the Hubble Space Telescope, ground telescopes, and X-ray telescopes that have occurred over time and linking them to our model using real physics. We have programmed the physical laws to simulate the dynamics of these events, including the colliding galaxies. If you look on that monitor, you can see our simulation of the colliding galaxies and compare it with the actual collision captured by the Hubble. You'll notice the simulation closely matches the actual event. We also have the ability to speed up the collision to see what the results will be in the distant future.

"You folks will never see this simulation anywhere else." Jason sat back in his chair and crossed his arms with pride. "They are nominating us for a Nobel Prize because of the macrophysics we have captured. What we have shown you should give you a better understanding of what our unique computer simulation is doing. You thought our universe was calm, serene, and quiet. But fortunately or not, we can't hear much because of the nature of empty space. Pressure waves and sound waves do not travel though space. Otherwise, we would hear more than a few loud bangs. Only electromagnetic waves, which include light rays, X-rays, and cosmic rays, exist in space."

"Good point, Jason," Austin said.

"Let's get back to our lab work here," Jason announced. "The sun's gargantuan turmoil is perceived on the earth only as a flickering, pleasant heat since it is dispersed over such a broad area. In other words, we don't even notice it. Thus, we sit here on any given starry night and perceive calm and serenity, but in reality, out there is enormous cataclysmic chaos. So much for the demonstration. What do you folks think?"

"Awesome," Tyresha said. "After you clobbered the analogy story this was excellent, Really!"

"Same here," one of the cute girls agreed. "That was breathtaking! Thanks for inviting us."

"Yes, it was breathtaking," Austin said. "Why don't you come over here and give me your phone numbers so we can contact you for future demonstrations?"

When Tyresha began to leave, Austin took notice. She nodded in his direction. "You guys are awesome. I thought you were all nerds, but this is really impressive." After she left the lab, she stuck her head back inside and smiled. "But will it give me my sixty dollars?"

Jason picked up a tablet and threw it at her. She chuckled and kicked it back into the room.

Tyresha walked down the hall to the office of a friend and told her what she had just seen. "Those guys in the lab are really sharp. But they don't have a clue about my real background."

"What do you mean, Tyresha?"

"For one thing, they would be stunned to know I got my job here because I was a science major at the junior college. This job is just a stepping-stone for me till I pursue my degree in physics next year. I was already familiar with most of what they were telling me. I just pulled their legs, is all."

"You go, girl! I know what you mean, girlfriend. They assume any stacked secretary from the ghetto doesn't know much."

"That's the way it is. But I just played with them. I don't fault them for making assumptions. It's just the negative perception that gets to me. Anyway, I like those guys, especially Austin."

Back in the lab, the other secretaries and visitors were blown away by the demonstration.

"Thank you for coming," Jason said. "I'm glad the demo helped. It's refreshing to have people appreciate what we're doing. Now we nerds need to get back to work. We still find some strange things happening out there that need our attention."

Later that day, Jason and his colleagues, in reviewing the data, continued to unearth disturbing information. They noticed that not every satellite was positioned where it should be. All the data indicated that the satellites orbiting the earth had shifted, along with some unexplained shifts in stellar locations. What they observed was, by all accounts, not possible. They couldn't accept or explain the results and threw up their hands in frustration.

Finally, they decided to reset their baseline data and begin reentering it into their model. They would redo the analysis the next day and correct their results. If the situation didn't change, science was in deep trouble.

19

Ocean Tsunami Detected

That afternoon, the director of the National Oceanic and Atmospheric Administration (NOAA) in Bethesda, Maryland, called an emergency meeting. His technical adviser, Dr. Strickland, had requested the meeting because he needed to divulge the astonishing news they had been shielding from the other staff members.

The overworked director received credit for running a well-organized bureau but was having a hard time keeping up with everything Dr. Strickland and his staff were doing. The director was under increased pressure to cut costs, which he had chosen to accommodate by reducing his staff size. He no longer had a press secretary or speechwriter and was forced to have Dr. Strickland present newsworthy items directly to the press and public. However, Dr. Strickland needed to be certain before releasing important information, which often resulted in unnecessary delays. But this time, he had requested that the director summon his staff, as Dr. Strickland had checked and double-checked his startling findings.

When the staff members arrived, Dr. Strickland took a deep breath before addressing them in the closed-door briefing room. "I want to thank you all for coming on such short notice. I don't know the best way to say this, but here goes. There has been a huge tsunami detected in the middle of the Pacific."

Ocean Tsunami Detected

The thin-framed Strickland, who never looked anyone in the eye, looked down and paused, collects his thoughts.

"What do you mean a huge tsunami? How huge?" one staff member asked.

"Here is what we know for sure. The satellite data we have just received indicates it is fifteen hundred miles long."

"What??" blurted another staff member. "Did you say fifteen hundred miles?"

Everyone in the room gasped.

"That's right. It stretches from Antarctica northward past the Easter Islands to the Galapagos Islands."

"That's not possible! It must be some type of ocean depression or tidal wave," the staff member said.

"No, it is continuous, and the entire wave is moving westward."

"How high is it?"

"Our preliminary data indicate it is over fifteen hundred feet high."

The room was wrapped in silence. Stunned and speechless, the staff members stared at one another in disbelief.

"You have got to be kidding! A tsunami that size would engulf entire islands. That dwarfs the one that hit Sri Lanka and Thailand in 2004!"

"You're right. It is gargantuan. We are not even sure it qualifies as a tsunami. Normally, a massive earthquake, a submarine landslide, or perhaps a volcanic eruption is at fault. But we have no recordings of any such major events. The only remotely related event is the widespread 4.1 earthquake."

"We have not seen any climate that forecasts anything like this," another member said.

"That is exactly what our instruments are showing," Dr. Strickland said, attempting to conceal his worry.

"Dr. Strickland, the largest recorded tsunami isn't nearly that large. This sounds like a catastrophe."

"You're right. The Hawaiian Islands are directly in its path. Strangely, the tsunami is moving slowly compared to the speed of a normal one. Typically, tsunamis move undetected in the deep ocean as fast as a jet liner. Fortunately, this surface wave is not like that."

"What do you think will happen next?"

"It looks like the Hawaiian Islands are on the verge of being overtaken by seawater."

"Wait a moment. This can't be true. What could have caused this? Maybe your data is incorrect. Nothing like this has ever been recorded," another staff member commented.

"Slow down, people," Dr. Strickland said. "This is exactly why we are calling this emergency meeting and why we have not notified the press. The data is correct. However, we are still missing data from one of our satellites."

"What do you mean, missing data? What's wrong?"

"To be honest, one of our weather satellites cannot be tracked or located."

"You mean an interruption in communication, not missing. Right?"

"No. I mean missing. In fact, the NASA satellite-tracking bureau says there are now three satellites missing. Gone, without a trace."

"Wait a minute," one frustrated staff member said. "Suddenly we are missing three satellites? And a mega-tsunami is about to annihilate Hawaii? Do you have any other bombshells you think might be of passing interest to us?"

"No, that's all we know at the moment. Anyway, we are repositioning the other weather satellites to view the Pacific to better understand these phenomena. We are as baffled by all this as you. The tsunami might be related to the worldwide earthquakes. We just don't know yet."

"If this is happening here, we must notify the Hawaiian Island authorities so they can begin evacuation proceedings," another staff member said.

"Evacuation?" Dr. Strickland said. "I agree with you about the notifications, but we suspect the tsunami may be even more widespread than initially thought. The wave is likely to draw in water from the shores of the entire western coast of the United States, but we have to double-check the data and get confirmation before we risk creating chaos."

"Holy hell!" the director said. "Dr. Strickland, we can't wait too long. We must recommend something now!"

"Sir, I know time is of the essence. That is exactly why I asked you to call this meeting. We are struggling with that as we speak. You need to take whatever measures are necessary while we continue to evaluate the data, but we must get confirmation before we panic the whole world. For the moment, I am providing a heads-up, but I am certain we have a catastrophe looming. The information will be confirmable soon."

"How soon?"

"I'd say by the end of the day we will know where we stand for certain. In fact, I recommend you go ahead and begin preparations to issue a category-five alert as soon as I receive confirmation of the data."

"Dr. Strickland, I don't think we should wait that long. If you are reasonably confident, let's begin alerting the presidential chain right now. Together we can face the heat of not having all the factual data on hand. Clearly, the president will want more input than just our opinions. The array of other astonishing and mysterious events will stun her. Our planet may be telling us something."

"Sir, I don't think we should contact the president until we double-check my data."

"Dr. Strickland, you had better stop making excuses and make the call right now!"

20

Tornado Tragedy

Britney and Manuela left the CSULB student union building and walked quickly to Manuela's car. They had not grasped the seriousness of the worldwide earthquake and unexplained sky scrolls, but their anxiety had intensified after the reports of more widespread sporadic lightning and tornadoes.

Britney and Manuela had been good friends since early in high school. Manuela was the only one who was aware of Britney's dilemma about going on tour with the band or continuing uninterrupted with her educational pursuits. As they walked, they noticed several science students in the parking lot looking up and pointing at the night sky. When Britney and Manuela looked up, they didn't see anything. So Manuela approached one of the male students and asked him what they were staring at.

"Look at those bright stars over there," he said. "Look where they're located."

"What?" Manuela shrugged her shoulders. "I'm looking, but I'm not seeing anything."

"Over there. Look closely. Notice their distance from the moon," he said.

"Okay. Why?"

"Just look at them for a moment."

"That's strange," Britney said. "It looks like all the constellations have shifted." She had had plenty of exposure in astronomy from her parents and Jason even though she was unsure of her expertise on the subject. Manuela, however, didn't have a clue as to what they were talking about.

"Exactly! The whole sky seems to have shifted. It's like nothing I've ever seen before."

"What the hell are you fools talking about?" Manuela said. "You weird science dudes are scaring the bejesus out of us again."

"Hey, baby, I don't know what is going on either," the student said. "This is some weird stuff. We're just trying to figure it out."

"This is too eerie for me," Manuela said as she stared at Britney, who seemed shaken up. "We don't need to worry about any of this technical stuff. Let's get out of here and let these little scientists worry about it, Britney."

Britney and Manuela made their way quickly to the car and drove away.

Four of the science students lingered looking skyward and noticed the sudden formation of clouds in the night sky overhead. The breeze started to pick up dramatically and soon turned into strong gusts. To their horror, the clouds suddenly yielded a spiraling funnel that touched down near the outskirts of the parking lot.

"Hey, dude, look at that!" one of the students shouted.

"Whoa! That is weird!" another student said. "It looks like a tornado to me!"

"Dude, we don't have tornadoes in southern California!"

The tornado savagely snatched several cars from the parking lot as the students stood paralyzed. No one had ever seen such a sight. As they watched in bewilderment, they didn't know where the cars had gone or if the tornado would start moving in their direction. It finally dawned on them that they had better seek shelter.

They began walking and then running, but it was too late. Screeching down from the dark sky came one of the cars, which slammed in the parking lot with a thunderous crash onto other cars, sending debris in all directions. Flying fragments struck some of the students. All those still in the area panicked and took off running.

The four horrified science students darted up the hill toward the student union building. Suddenly, two cars crashed down near them, crushing one of the four students. Sheer bedlam ensued as screaming students scrambled for cover.

The tornado subsided as quickly as it had appeared. The clouds dissipated, leaving devastation and carnage across the entire campus.

At the same time, the same type of clouds suddenly began forming around the landmark ARCO Tower in downtown Los Angeles.

21

Space Station Docking

Jason returned home that evening anxious to watch the docking of the Falcon 15 with the space station on TV. It was almost time, but Britney was not yet home.

Their father, Walter, and his crewmate, Larry, were on board the Falcon 15 looking forward to the docking process. Their mother, Dorothy, was busy at work in the International Space Station on her redirected task of collecting spectrographic information on the sky scrolls. She was confident that they had things in hand.

As the Falcon approached the space station, NASA scientists noticed that the two massive structures were not properly aligned for docking, so NASA officials decided to put a temporary hold on the docking process until they could determine what was wrong. All the while, the media continued to bombard them about the unexplained breaking news.

Jason suspected that the misalignment was related to the phenomena associated with the weather satellites that his team had been examining. In other words, all calibration adjustments using the stars appeared to be a little off. He was also concerned about the bizarre orchestration of the coincidental and mysterious atmospheric events.

Once the hold was lifted, NASA continued with the scheduled docking, but switched from automatic mode to manual mode. This was facilitated using a tether device to tweak the two massive

Space Station Docking

vehicles into final alignment. This rare backup approach was working well, and the Falcon capsule and space station appeared to be on the road to a successful alignment and docking.

The transfer involved sending one of the crew members from the capsule into the space station and replacing him or her with an astronaut from the space station. The hatch opened, and Walter and Larry prepared to traverse from the capsule to the space station. So far, everything was going according to plan.

The process was this: The crew members would transfer through the narrow, tube-like hatchway that protruded into the space station. They would move into the transfer hatch, and the capsule door behind them would close. They would then pressurize and perform a systems check on the hatch tube. They, then, would walk to the space station side. After checking everything out, the crew members would exit safely into the space station, and the hatch door would close behind them. The two crew members leaving the space station also performed this three-step process in reverse.

As Larry and Walter entered the transfer compartment, the door behind them closed as expected. However, Larry noticed some jittering in the thin, taut tether line used to cinch the docking process. He pointed this out to the NASA officials over the communication line. They indicated this was to be expected and reassured him that the docking process was stable and going well.

Just as the final pressurization process was completed, Larry and Walter, still inside the transfer compartment, felt a sharp jolt. Instantly, the tethered cable snapped and whipped through the transfer compartment, decapitating Larry. As his torso flailed around, his spurting blood coated the instrument panel and video camera lens. Walter, shocked and horrified, was also doused with blood. As Larry's torso convulsed and his head spun weightless, the eyes still blinking, Walter screamed and collapsed in a state of shock.

The captain of the space station heard the commotion and asked what was going on, as the blood-covered video camera lens hid the gruesome scene. Silence met his repeated attempts to communicate with Larry and Walter. Larry's erratic vital signs suggested that he had inadvertently tampered with his electrical body leads. As a result, the captain and Dorothy decided to momentarily halt opening the space station door.

Meanwhile, the capsule and station continued to undock automatically since everything was going well from the capsule side. The two vehicles undocked and began moving away from each other while Dorothy, the captain, the other crewmembers, and the sealed transfer compartment moved with the station. For a moment, there was a glitch in communication within the transfer compartment.

Had they known what awaited them, Dorothy and the captain would not have waited until the pressure in the transfer compartment stabilized before opening the door to see what had caused the commotion and the clouded video camera.

22

Weather Calamity

Two mornings later, at NOAA, the national weather bureau, additional satellite data finally confirmed that an immense tsunami had occurred in the middle of the Pacific Ocean. The photos showed that the entire ocean had experienced a wide parting along a line running north to south, displacing unimaginable amounts of water. This water had begun propagating in a westerly direction with the likelihood it would engulf every island in its path and collide with the shores of the western Pacific Rim countries within days in the form of a tsunami of massive proportions. Dr. Strickland immediately issued the highest emergency alert level possible to all affected nations and the public.

With the news of the impending water-borne disaster, the eerie sky scrolls, and the sporadic tornadoes and lightning, near pandemonium began erupting all over the world. Many people stayed home from work and assembled in churches, mosques, synagogues, and temples. However, most of the public did not accept the scientific speculations, including NASA's flawed explanation of the worldwide geologic, atmospheric, and oceanic phenomena. By all appearances, years of unchecked scientific exploits and utter disregard for nature were beginning to backfire.

23

Church Sermon

The public was now seeking spiritual guidance and protection. Though not all were aware of the scale of the tsunami poised to wreak havoc, they were keenly aware of the dreadful environmental conditions plaguing their lives. The increasing sporadic lightning strikes and the surprise tornadoes created a climate of fear. However, if the public did become aware of the scope of the calamity brewing, desperation and pandemonium would rule the day.

Britney, Jason, and Manuela huddled together at their nondenominational church listening to the sermon. Jason's family often attended this church together, and sometimes Austin tagged along. However, when Austin and Manuela were together, they usually clashed since Austin didn't trust women and Manuela didn't trust men and strongly advocated women's rights views to Austin's face. Manuela also thought that Austin's contribution toward Jason's celestial body research at UCLA was inadequate. She believed he was riding on Jason's coattails. Her distrust of Austin further compounded her criticism of scientific research in general and specifically its far-fetched technical theories about the universe.

In a special radio broadcast, their minister espoused that God had a plan, and that Scripture indicated we would receive a sign about the revelation.

"However, there is some good news here," he said. "These unexplained events are pulling all of us together throughout the world as we search for answers. Is the world's climate changing? Is the earth going to explode? Is God sending us a message of Armageddon? Are our sins coming to a head causing God's anger?. None of us have the answers to these questions, but we must have faith that God's will is being done."

Because of the worldwide fear, their minister had consulted with other religious leaders around the world. Based on these discussions, as well as his previous world religion studies, he had put together some provocative notes. Britney, Jason, and Manuela were especially attentive, partly out of respect for the religious institution and partly because they were open to any explanations that might help quell the fear and uncertainty that plagued everyone.

Manuela whispered to Britney just before the sermon, "I believe Mother Nature is behind all of this, and there is a divine plan in the works. I reject these bizarre scientific explanations. This is why I migrated from the Catholic Church to a nondenominational one."

"I would like to share some thoughts with you," their minister continued, "with the hope they will help us all understand why God is so affecting the planet. These thoughts are not dependent on our individual moral or religious beliefs. To put things into perspective, let me talk about world religion for a moment."

Britney, Jason, and Manuela began to fidget, as they each had their own beliefs in spiritual matters.

"Religion is a practice and refers to the fear or awe one feels in the presence of a spirit or a god," he began. "Technology has made the world a neighborhood but not a brotherhood. We can push a few buttons or click a mouse and connect instantly with someone continents away. While living in a free society, many of us lock our doors, engage our security systems, and watch nightly news reports of kidnappings, murders, drive-by shootings, hate crimes, drug deals gone bad, and international terrorism. Graphic pictures of bloodshed bombard our senses and assault our respect for the sanctity of human life. We long for a system of beliefs and values that will help us cope with injustice, affirm human dignity, instill hope for a brighter future, and deal with fear and the unknown.

"In Western cultures, we define religion by a set of beliefs having to do with God through which one learns about a moral system. Although this definition contains elements documented within many of the religions of the world, it cannot do justice to them all. Most religions that have existed on Earth have been more concerned with humanity's proper relationship to gods, demons, and spirits rather than with ethical relationships among people.

"There is no significant example in history of a society successfully maintaining moral life without the aid of religion. Today, we face a challenge that may rival that of Noah and his ark or Moses crossing the Red Sea. Whatever happens, we must trust that God's will is being done. Some of you may not believe this, but we need not be alone. I would like to quote a passage from Psalm 23 that may help all of us through these trying times: 'Yea, though I walk through the valley of the shadow of death, I will fear no evil: for thou art with me . . .'"

Manuela peered at Britney to see if her fears had been allayed or only increased.

Church Sermon

"It sounds to me that he is suggesting we just wait and see what happens," Britney said as she toyed with one of her gaudy earrings.

Jason interpreted the sermon differently. "He's suggesting that everyone should just sit back and let nature take her course." Annoyed, he stood and slithered out of the church.

24

Lightning Havoc

On Monday morning, one week after the worldwide earthquake, Manuela drove to CSULB at Britney's invitation to hear her musical performance. It did not go well because of poor attendance and ended abruptly. They left together, discussing the concert and their uneasiness about the bizarre events of the previous Thursday.

They scurried through the partially vacated and cordoned-off campus. Police and security had roped off the tornado-damaged section of the campus and were barring trespassers. Everywhere they turned, people were talking about the incredible events of the past week.

The girls walked aggressively toward the temporary parking lot. They were nervous and frightened about walking out in the open, thinking that anything could happen at any moment. Suddenly, Britney stopped, threw up her arms, knocking one of her earrings to the ground, and started weeping.

"What is happening to us?" she blurted out. "Everyone has their own opinion about what's causing all this crap. But nobody knows what *is* happening! I can't live like this!"

"Come on, Britney," Manuela said. "Crazy things are happening. Be comforted by at least one of the things the minister said yesterday. People are overreacting. Nature goes through these phases ever so often."

"I don't think that the rest of the world and I are overreacting, Manuela. This is much more than curiosity. I believe that if some clear and convincing explanation is not forthcoming soon, things will be out of control."

"Britney, don't get caught up in all that hype. I heard on the news that some paranoid people in Europe have been hoarding food. That's so ridiculous. Our scientists will have answers for all of this soon enough."

"I hope so, before people start breaking into stores and stealing supplies like those crazy riot people."

"Come on, Britney, people are just overreacting. Do you realize that the people in the United States have been uneasy since the bombing of the World Trade Center in New York?"

"You're right about that. I have heard some people speculating that all of this is some sophisticated terrorist plot."

"Everyone is just freaking out." Manuela held both arms in the air with her fists clenched. "Just a few minutes ago, when we were in the music hall over there, some guys were saying the earth's core is becoming unstable and is about to explode. That ridiculous talk bothers the hell out of me!"

"It doesn't surprise me, though, that people are really scared."Britney added.

"But I still don't know why people are panicking so much. Nothing tangible has hurt us yet. Be patient; our scientists will solve these mysteries. I am certain that everything will be explained in due time."

"How can you be so damned calm?" Britney said. "The entire earth has experienced a sudden, tremendous jolt. Is it getting ready to explode? Have the earth's guts become unstable? What the hell are those scrolls all about? Why is there lightning all over the place? Is there some damn ozone hole or leak? Have we polluted the earth so much that it is about to split wide open? Talk about gloom and doom. I am damned scared, girlfriend! Your scientists caused this mess. They have gone against nature and our Maker." Britney collapsed to the ground, sobbing and pounding her fists on the grass.

Manuela put her arms around Britney to comfort her. "Okay, okay, take it easy, Britney. I am just as concerned, and maybe you're right. Okay, I'm scared, too, but there is one good thing. I see many people starting to pray now."

Manuela held Britney for a few moments. Once Britney was back to herself, they walked to Manuela's car that was parked in the lower campus parking lot. It was then that several bolts of lightning struck the ground behind them. The noise was deafening, and the light was blinding. They both screamed and dropped to the ground when the building they were standing near got slammed by a huge bolt. The lightning roared down the side of the building, hurling a multitude of heavy bricks onto the scrambling students below. The bricks pummeled several students while others scattered in all directions.

"Oh my God!" Britney screamed. "Let's get out of here!"

They ran down the hill toward the gym building. Several students running from the music building joined them in their sprint. Suddenly, another bolt of lightning came out of nowhere and struck a slow-running, dumpy student. The student tumbled to the ground and flipped over several times. Horrified,

Lightning Havoc

Manuela's foot slammed into a low-lying brick walkway adjacent to the sidewalk and tumbled to the ground, her long, black hair wrapping around her face and her headband flying off. Her ankle was severely twisted, and she was unable to move.

Britney continued her dash to the gym building, unaware of Manuela's predicament. When she finally saw Manuela struggling to crawl to the gym, Britney motioned to some nearby students to go back with her and bring Manuela inside, but they ignored her and continued fleeing. So Britney ran back alone. She grabbed Manuela and tried in vain to help her up. Two male students saw the struggling girls and stopped to help. Within minutes, they had all made it into the gym, where they crouched on the floor and huddled together.

The lightning finally stopped. But anxiety reigned throughout the gym as everyone tried to figure out what had caused lightning in the early evening without any warning or indication of a storm anywhere.

The students were unaware that these strange phenomena were occurring throughout the country and even the world.

25

Public Pressure

President Heather Rachael Clemson was asleep with her husband, Bailey, in the White House the following morning. It was Tuesday, eight days since the initial earthquake. Bailey had been one of the more well known figures in the political arena on his own merits even before his wife became president. He was well-respected and experienced in national and international matters.

Heather was awakened at 3:00 a.m. by her automated breaking news alert system. She quickly jumped out of bed to review the news on her system monitor. She dashed to the adjoining room to her computer on the mahogany table which had automatically switched on and was starting to boot up. Heather gazed at the wall where pictures depicting important events and occasions from her illustrious career were displayed. Most of her staff viewed her as a personable, respectful, soft-spoken extremely competent woman. She worked effectively with her scrupulously chosen assistants and advisory staff. A relentless drive for solving complicated domestic and international issues was her forte. Attractive with piercing blue eyes, her fashion sense was impeccable. TV and pictures did not do her justice causing many in her path to be caught off guard by her beauty. She could be stern while remaining mild-mannered with a smile. On those rare occasions when she would lose her temper, she could unleash the wrath of hell. Many people believed her presidency demonstrated bold and unselfish leadership. She always made the most well thought out decisions for the country, even if it meant sacrificing some of her power and control. Her views were more like those of Gandhi than those of a typical Western president.

Heather sat down in her elegant ergonomic chair and scanned the breaking news.

"Honey, you won't believe this." She grimaced, jerked backed in her chair, and accidentally knocked some papers from the table. "The alert is that 98 percent of my incoming e-mail is focused on the sky scrolls, earthquake, and sporadic lightning throughout the country. What is going on?"

"That doesn't surprise me," Bailey replied in a low voice, still partially asleep. "Everyone is talking about them. Your staff has sheltered you far too much from the public fear and anxiety."

"Bailey, I am quite aware of all the concerns," Heather replied. "My people are on top of them."

"Bullshit. None of your people are on top this, honey." Bailey sat up quickly, causing the bedspread to slide to his waist, exposing his bare chest and thinning gray hair. Heather continued typing at her ninety-word-per-minute pace, reviewing the barrage of news reports, and scanning several briefings focusing on the mysterious earth phenomena and tragic events of the previous day. Of particular interest to her was the description of spontaneous lightning striking a university campus, killing two and injuring several other students. A similar phenomenon occurred in downtown Los Angeles, where the ARCO Tower was struck, raining debris on bystanders.

"Honey, listen to this," she said. "The frequency of lightning strikes is reaching alarming levels. Each year, over one thousand people are struck in the United States by lightning and kills close to one hundred of them. During the past six days, there have been over 120 deaths from random lightning, and information is still coming in.

"Listen to what's on my other SC (secured communiqué)," she continued. "A suspected tsunami ocean swell is approaching the Hawaiian Islands. These are the most horrifying news briefs I have ever seen. This situation is making me feel numb. I must drop everything and devote all of my attention to this."

"Dah!" Bailey responded.

Heather's face grew pale as she continued scanning the barrage of breaking news events. Her voice was tinged with alarm as she read aloud to Bailey. "I'm not sure what the heck to do. What in God's name could be going on?"

"Heather, it sounds like you had better move this situation to the top of your priority list. Why don't you convene a special meeting with your technical staff ASAP and brainstorm what should be done?"

"I think you're right." She picked up the phone, contacted her twenty-four-hour, on-call planning secretary, and said, "I want to cancel all of my morning activities and convene a special meeting in the Oval Office tomorrow morning with my technical staff. I want to include one of my specialists from the geology department on earthquakes and an expert on volcanoes."

"What about Homeland Security?" Bailey added.

"Include our Homeland Security officer and an expert from the National Oceanic and Atmospheric Administration. Tell them not to worry about a big presentation. Our need is to brainstorm solutions ASAP. We have to figure out what the heck is going on with our planet so each can weigh in on a direction on which each should place their efforts"

"I will get right on it," the secretary said. She frantically but methodically began making all the appropriate calls.

By that evening, the news of the emergency White House meeting had leaked. Rumors began to swirl. One evening tabloid report speculated that the volcanic eruptions had created the worldwide earthquake and that geologists were attempting to determine if the earth was splitting open down the middle. Another report claimed the Homeland Security adviser was going to the White House to divulge that terrorists were behind the mysterious phenomena and that the United States was preparing to negotiate a truce with them.

Despite the wild speculations, the meeting moved forward and the staff dignitaries convened in the Oval Office the following morning. President Clemson looked around at her staff gathered in the conference room and made a mental note of who was there. She allowed her husband to sit in on this unique meeting because of his insight and extensive international experience and influence.

"As you all know by now, we find ourselves in a grave situation," she said. "I am not looking for a political approach. I need concrete solutions, and I need them now. The public is demanding answers and explanations. There has been too much pain, suffering, and grief for parents and loved ones already. What does all this mean? To be frank, I am not only extremely concerned, but am frightened."

The meeting room was silent.

"First, I want to thank each of you for coming here this morning on such short notice," she continued. "I don't think any one of you has all the answers, but I want us to take a few minutes to discuss what is going on with our world. Perhaps we can then move into a more structured agenda. Is this okay with everyone?"

Everybody nodded. She looked at Thomas James, the head of the Joint Chiefs of Staff, and requested that he start the discussion. "Tom, what do you think is happening?"

He looked up. "I think we are getting a grip on things," he said in a high-pitched, unsteady voice. He looked at his notes. "The army, navy, and air force all have their resources on high alert. We are considering sending a directive to suspend all air traffic within a one-hundred-mile zone beneath the so-called sky scrolls as a precautionary measure. NASA is redirecting two of its earth-orbiting satellites to measure the physical contents of these scrolls. The office of geological exploration is compiling data on the movement of the earth's tectonics plates. NASA is proposing canceling all near-term space-flight launches, and we plan to redirect and reexamine all activities on the space station to ensure the crew members are safe."

"Tom, you are doing many things," Bailey said. "But what the hell does all that mean? What is happening? Is the earth getting ready to explode? Is our atmosphere ready to evaporate? What is going on? You had better get your act together and have your top people give us some answers."

"That's enough, Bailey," Heather interjected. "This is my meeting. Another outburst like that and I will have to ask you to leave."

Heather paused as Bailey sat back in his seat, crossed his arms, and dropped his head. Since she had never served in the military, Heather did not want to show any weakness in her military leadership. She wanted to demonstrate that she was in control of the situation and would not crack under the stress of this combat-like situation.

"Let's start again," she said. "What do you think is happening to us from a geological point of view?" she asked Dr. Tom Mauls of the Department of Geology.

"I don't think the events are all necessarily related," Dr. Mauls replied with some degree of confidence. "I think the earthquakes are due to tectonic plate movement, and the volcanoes and massive lava flows are a result of this movement. So far, our data indicates the volcanic eruptions may have created the earthquake, and our geologists are assuring us the earth is not splitting open. All of our geological parameters show nothing unusual now."

"Thanks, I hope to God that you are right," Heather said.

"What about you, Mr. Stevenson?" she asked the secretary of Homeland Security. "Do you think there is any validity to the terrorist theory?"

"Absolutely and unequivocally no," he replied in a stern voice. "There is no terrorist group in the world that could have caused any of these events. All the resources of the United States government could not have created any of these phenomena let alone some underfunded, clandestine terrorist organization."

"I tend to agree," Heather said, smiling at his confidence. "However, I think I heard similar comments from some of your colleagues in the past. Forming the Homeland Security Department was a result of the 911 terrorist attacks in which officials claimed nothing like that could ever happen on our soil. Nor could a passenger plane with 270 people onboard simply vanish without a trace. Similarly, you might recall the French made the mistake of downplaying the advances of Hitler's organization. Therefore, I think we need to keep an open mind. In any event, I'll get back to you.

"Mr. Lewis, I hope and pray that we are getting closer to explaining the sky scrolls. What is your thinking now?"

"Ladies and gentlemen, the Department of Interior is aggressively working on this. We believe there is a scientific explanation for these phenomena. We are working feverishly on this, but we don't have a clear understanding yet."

"Well, Mr. Lewis, would you give us your best hypothesis?"

"I believe it has nothing to do with the internal stability of the earth but is related to an external phenomenon."

"Do you mean we are being attacked by aliens?" Bailey said. "Or do you mean this external phenomenon is the hand of God slapping us around?"

"Bailey, I am warning you again." Heather gave him a stern look. Despite her admonitions, she was actually thankful that Bailey was there to listen with a critical ear. She just wanted her staff to know she was not going to put up with his outbursts. "Please continue, Mr. Lewis. Ignore Bailey."

"I believe it has something to do with another celestial body, such as a meteor or comet or perhaps the moon. But that is only my personal opinion. We will have some facts and data soon to share with you."

"Thanks for the input," Heather said. "But I don't know where to go with that at the moment. Let's hear from NASA. Mr. Braxton, what do you think?"

The stocky, well-groomed director leaned forward, glanced at his Rolex watch, and said, "Good day, Madam President. My comments will only take a minute. You can refer to me as *Doctor* Braxton. As you may or may not know, my entire staff here and those from Down Under in Australia have been aggressively investigating this bloody mess. Frankly, boss, I think your staff here is blowing smoke up your ass!"

Heather's eyes flew open, wondering if she had heard him right. The startled expressions of her staff confirmed that she had.

"What did you say?" she said.

"That's right, mate," he said, gritting his teeth. "We are faced with a world catastrophe, the likes of which mankind has never experienced. It is possible that doomsday or Armageddon is upon us. We had better collect every possible resource we have. We are not a people who simply give up, fall to our knees, and pray that things will get better. We need to determine our own fate. We need to work not only as a nation but also as a world unit to muster all the knowledge and resources possible to address these phenomena. All signs indicate our planet is in serious trouble."

Heather sat motionless, and none of her staff budged.

"I respect your candor," Heather said. "You certainly have my attention. If all of you think this is about doomsday, we must do everything to protect our citizens and our planet."

Those in the room exchanged glances and began nodding in agreement with Dr. Braxton's comments.

"Allow me to continue," Dr. Braxton requested. "There are many groups, organizations, governments, and nations independently investigating these events. We need to change this. We need to unite our efforts."

"I agree," the president replied. "Perhaps we should convene the presidents, prime ministers, and leaders of the major countries."

"I agree, Madam President," said Dr. Braxton. "I think it is best to engage these chaps and include the UN in this endeavor since it affects the whole world."

"You're right," Heather said. "We definitely need to involve all nations, and I believe the UN is the best way to do that."

"Let me add something," interjected Mr. Stevenson. "The UN Security Council is set up to deal with this situation. Its main charter is maintaining international peace and security. We certainly need to address international security."

"Well, then," Heather uttered, "let's contact Secretary-General Bay Loy from China and have her convene a worldwide summit to get to the bottom of these events. We should also include the leading scientists and astronomers."

"Madam President, I think that is an excellent idea," Dr. Braxton remarked. "We Aussies have considered doing something locally, but your worldwide plan sounds better. And what's more, it sounds correct. I can help pull this meeting off within the next few days with your backing."

"You have my support."

"Thank you, Madam President. I suggest we convene in Southern California where many of the scientists are doing much of the research that will bear on the summit."

"I don't care where you have it, but let's get the appropriate people together somewhere and let the public know what we are doing."

"Yes, we can do that. I will ensure that the summit begins within seventy-two hours. It will take place at the Vandenberg Air Force base rather than the UN building in New York because of access to critical research information in that region of the United States."

"Agreed. I will ask our ambassador to the UN, Amy Davis, to work with UN Secretary-General Loy. They can create the proper protocol and take the lead in organizing and executing the summit. I will commit whatever resources are needed."

"Madam President, I suggest we invite representatives from selected countries to be part of the main discussion session and perhaps simultaneously teleconference with the space station. We should begin with a presentation of the facts so everyone is on the same page. This information should be distributed before the meeting, and we could then allow one representative from key disciplines to present their findings and hypotheses."

"Don't forget," Bailey intervened again, "that we should include representatives from the religious sectors as well."

"Sounds great," Heather said. "I think that all of this is an excellent plan, and I also detect the right sense of urgency here. Does everyone agree?"

Everyone in the room nodded.

"Okay, let's roll." Heather stood up and raised her arms with clenched fists as if she had just served an ace on the tennis court.

26

Oaxaca Rescue

Early Thursday afternoon, Jason heard about the emergency summit called to commence on Friday. It was now over a week and a half since the initial earthquake and the sky scrolls first appeared. The Pacific Rim countries and the Hawaiian Islands were scrambling to deal with the impending mega-tsunami.

Jason had been working nonstop focusing on incorporating the piece of celestial wind effects that Thatcher had failed to discuss. He now had some breakthrough results he needed to share. He was still unaware of Professor Thatcher's evil deed and Austin's vengeful actions.

Furthermore, Jason had been desperately trying to contact Jane. He had called the hotel in Oaxaca where the students were staying but was not able to get through. Apparently, the phone lines that transmit out of the country were jamming due to damage and the extraordinarily high call volume. When he tried Jane's cell phone, he found that the roaming feature was not working.

After hours attempting to contact her, Jason was desperate. He debated between staying put and disclosing his breakthrough findings or dashing to Mexico to find Jane. He wanted to ask her to marry him and give her the engagement ring he had bought right after she had left. He was planning to propose to her upon her return from Mexico, but given the fact that the FAA was considering canceling or suspending flights due to the sporadic lightning in regions near the sky scrolls, he made an abrupt change in plans. Jason headed to the LAX airport.

At the United Airlines ticket counter, Jason was informed that most of their flights out of the country were suspended. When he complained that Mexico was not exactly out of the country, another customer, overhearing the argument, whispered to Jason that they were still booking flights to Mexico for diplomats and their immediate families on AeroWest. He immediately raced over to AeroWest, where there was a long queue of people. He got in line and, while he was waiting, concocted a story to convince them to allow him to fly to Mexico.

As chance would have it, a businessman standing behind Jason grew annoyed with the long wait and stormed up to the counter to confront the ticket agent. "Excuse me," he said. "I need service now! It is very important that I get on this flight."

The clerk looked up at him. "Sir, I understand your urgency," she said. "However, you have to wait in line. We will get to you as soon as possible."

"Do you know who I am?" he yelled as he slammed his e-ticket on the counter, startling the agent. "I need help now!"

The agent picked up the airport intercom. "May I have your attention, please? May I have your attention?" Her voice echoed throughout the entire airport, and the irate businessman looked on, surprised. "There is a gentleman at ticket counter twenty-three who does not know who he is. He says he needs help now. Can someone please come help him identify himself?"

Stunned and embarrassed, the businessman slowly backed away from the counter. Everyone stared at him and snickered as he slithered alongside the counters and disappeared down the hall.

Jason grew concerned that he would not be able to get a flight out to Mexico. However, after telling his story to the agent, she sensed his sincerity and urgency and issued him a ticket to board the flight.

Jason arrived in Mexico to find many of the airport services had been suspended, including cabs, the baggage carousel, and information desks. He desperately searched outside the airport and stopped a local resident.

"Sir, I need your help. Can you please drive me to the University of Oaxaca campus dorms?"

"Me no speak English well."

"Thanks. Okay, I'll pay you. Please take me."

After a tumultuous high-speed trip over unpaved roads, they finally arrived at the campus. Jason jumped out of the car, paid the driver and hiked to the dormitories, racing from room to room in search of Jane. Finally, one of the resident students indicated that the UCLA team had already departed for the United States. Jason queried the student in broken Spanish, pleading with him to recall the airline and flight number. The harassed student indicated they had taken a private jet that departed from a local landing strip.

Even more desperate now, Jason coerced another local resident to take him to the landing strip. They piled into his rickety pickup and headed down the pothole-ridden streets.

"Ándele, pronto, hurry up!" Jason yelled, gesturing for the driver to speed things up. He knew his time was running out.

The driver raced through a few neighborhoods, swerving to avoid animals and buildings. Suddenly, there was a loud bang, and the car veered out of control. It swerved to the right and then to the left before ending up on the side of the road in a shanty neighborhood. Jason and the driver jumped out to survey the damage.

"Donde esta su spare tire!" Jason yelled, gesturing at the blown tire.

"No tango la llanta sobrante. Aye, aye, no have tire," the driver confessed as he held out both hands.

Jason was stranded in the middle of nowhere, in another country, without a clue where he was. All seemed lost. He might never see Jane again, and he had no chance of returning to the summit to reveal his findings. He slumped against the car.

Suddenly, he heard a recognizable humming and buzzing sound. He looked up and saw a small airplane climbing into the sky. The airport must be nearby! But how could he possibly get there? Jason looked around and saw a rickety bicycle leaning against a shack. Without hesitation, he grabbed it and headed in the direction he believed the airplane had ascended from. After several wrong turns and dead ends, he finally reached the airstrip.

Seeing a school bus in the distance near the runway with people milling around, Jason jumped off the bike and climbed over the fence. He ran toward the bus in hope that he would recognize someone.

Immediately, he saw Jane's faculty guardian. Frantically, Jason indicated he was searching for Jane, and the guardian pointed across the airstrip. Jason sprinted toward her, and when Jane saw someone running and realized who it was, she dropped her bags and ran to greet him. They raced to each other and embraced as both their ponytails wrapped around their faces.

"I thought I would never see you again!" Jane said.

"I love you, Jane," Jason cried. He dropped to his knees, pulled out the tiny box, opened it, and removed the engagement ring. "Will you marry me?"

The other students, including the guardian and Luther, looked on with surprise and envy.

"Yes, yes, yes!" Jane cried. "I love you so much!"

They embraced for several moments, and many of the students began clapping.

"I am happy to see Jason propose to this high yellow sistah. I think they will make a good mixed couple," one student whispered to his friend.

"What? I didn't think she was a sistah, but she does sort of look like a long haired, young Halle Berry," the other student whispered back.

"Whatever, it's hard to tell," the first student said.

Eventually, things settled down, and Jason and Jane resumed thinking about the predicament they all now faced.

"Why haven't you guys left yet?" Jason asked.

"The pilots were so fearful of the sky scrolls they abandoned the aircraft, and we have no way out," Jane said.

"What? I have to get back to the United States and join the summit!" Jason said. "I have been running my computer program with the latest data, and I think I know what is behind these mysterious occurrences."

"What are you talking about?"

"It's a complicated story, but I've gotta get back. I don't know exactly how, but the latest information we put into my model confirms something incredible has happened to us."

"What do you mean, to us?"

"I am talking about the entire planet."

"I still don't understand," Jane said.

"I'll explain it to you later."

When Jason realized they were stuck in Oaxaca, he began to cerebrate. He had no way to communicate to the summit, let alone return there and disclose his findings, but he had to find a way. Grabbing Jane's hand, he dashed toward the airport terminal. However, after a few steps, he stopped in his tracks and looked backed at the abandoned airplane on the runway. Then, holding Jane's hand firmly, he walked toward it.

Jason began scrutinizing the airplane. He recognized it as one his father had flown many times. He recalled sitting with his father and the times he had allowed Jason to ride in the copilot's seat. On several occasions, his father turned over the controls to Jason.

"I think I can fly this plane," Jason said. His meager experience of long ago was the sole basis of his confidence.

"You mean fly this airplane?" the guardian uttered with astonishment.

"My flying lessons were in a plane similar to this one."

"No way, Jason! We need to stay put," the guardian insisted.

"I can't," Jason stated defiantly. "I'm just going to take a look."

"I don't think that's a good idea. You need to stay put!"

Despite the guardian's insistence, Jason and Jane went aboard, and Jason began examining the cockpit. He recognized the controls and noted that the plane had a full tank of gas. He also saw an operating manual. He hesitated for a moment and then started the engine.

"Jason, get out of there!" the guardian yelled. "Stop tampering with the plane!"

Jane looked at the guardian, hesitated for a moment, and climbed aboard.

Jason told Jane to tell the others he was confident he could fly the plane and was going back to the United States.

"Are you sure you can do this?" Jane said.

"I know I can, with your help."

"*My* help?"

"Yes. *Your* help."

"Okay, Jason, I have faith in you," Jane said, standing behind him and putting her hands on his shoulders.

"Tell them I've flown this type of plane before, and I'm certain that I will have no problems."

Jane studied Jason for a moment and then left to make the announcement.

Two students hesitated but then cautiously came aboard.

Luther, watching this, did not want to be stranded in Mexico, so he pleaded with Jason. "Say, man. I need to get back to UCLA, too, and I hope you can do this."

"It's up to you, dude, if you want to take a chance. But you have to do what I say."

"Yeah, okay."

Jason yielded to the pleas and allowed Luther and two others to come aboard. They quickly did so despite the guardian's rants.

Oaxaca Rescue

"You guys get back down out of that plane! Are you crazy? I'm in charge here!"

They closed the door as the guardian continued screaming at them not to go.

Jason taxied the plane down the runway before bringing it to an abrupt stop, stunning and terrifying the students. "I forgot something," Jason said. "To be honest, I forgot to set some of the controls. Besides, I need to do a test run." This scared the hell out of his passengers.

Jason turned the plane around and returned to the opposite end of the runway. He asked Jane to open the control manual, find the section on takeoff, and sit next to him in the copilot's seat.

After skimming the manual, Jane found the section but skipped past the "Introductory Cautions" section on Special Conditions. This was a big mistake. If she had read through it, she would have noticed that the airport was situated in an area in which severe gusty crosswinds were likely to occur. The air traffic control personnel always suspended flights during such conditions. However, because of the global instability, all flights were suspended anyway and the air traffic control personnel were not on station. Unaware of this, Jason once again raced down the runway, the plane lifted off, and they were airborne. As they ascended, Jane reviewed the section on communication, and they figured out how to use the radio to communicate with the control towers.

Everything was going well until a wind gust suddenly jerked the airplane, causing it to start to roll over. Jason attempted to roll the plane in the opposite direction to level it out. Fortunately, this was just what the instruction manual prescribed, and the plane stabilized. The students, petrified after this maneuvering, clung to their seats.

A fearful Luther wandered to the back of the plane, saying, "I'm going to check things out back here." He noticed the emergency equipment and saw that there were only two parachutes. After thinking for a moment, he took one of the parachutes and hid it under one of the rear seat compartments.

During the flight, Jason was able to explain more of the details of his research findings to Jane. Astonished, she indicated that she believed Jason's theory was correct, adding she would stand by him. She felt he needed to contact the highest authority possible to explain his findings. Even contacting President Clemson was not out of the question, as Jane believed that Jason's findings warranted attention at the executive level.

As they approached Los Angeles, they managed to establish communication with the ground.

"This is Jane. We are coming in from Oaxaca, Mexico. We need help landing this plane."
Jane then explained the enormity of the situation and convinced the ground communication to put them through to UCLA.

After ground communication patched them through to Jason's research lab, Jason learned that they had verified his model's incredible predictions and agreed that the data supported his bizarre theory. One of the research students told Jason that no one knew he had developed a plausible explanation for some of the mysterious phenomena, and that the technical information conclusively supported it.

"You probably don't know about Professor Stein," one student said. "He was personally invited by the UN Security Council to participate in the California summit as a key technical member on the main panel. He has been desperately looking for you, hoping you would join him at the summit. Professor

Stein would much rather have you present these findings because you have the best understanding and the backup information."

"Jason asked if they could get in touch with Professor Stein."

"He told us that if we found you to tell you to go directly to the summit, and he would try to have all the roadblocks removed.", the lab student said.

"Jane, I think you should do just that."

"Jason, we will try to call ahead to the US ambassador to let the United Nations know you're coming with some breakthrough information.", the lab student said.

Jason agreed and was determined to land at Vandenberg Air Force Base. However, the other students onboard did not want any part of it.

"Hey, dude, you need to land this plane now! You can do your little talk later," Luther insisted.

The other passengers adamantly demanded that Jason land at LAX. But he stood his ground and suggested they not piss him off while he was flying the plane.

"Jane, tell ground control that we will need help communicating to the control tower and the air traffic controllers at Vandenberg if they are still around."

Jane opened the operating manual. Working together—and with the help of ground control—Jason and Jane managed to reach Vandenberg Air Force Base. After a harrowing approach, Jason managed to land the plane safely.

Jason, Jane and the other students quickly exited the plane and darted away to secure transportation to the summit. As they ran down the runway, Jane suddenly fell to her knees and started vomiting.

Jason stopped running and went back to her aid. "What's wrong?"

"I'll be all right," she said. "I have felt sick ever since the morning after I stayed in the hotel with Luther. I think I'm okay now." At that point, Luther arrived by her side. She stared at him. "It's probably from the accidental overdose of Valium I took."

"What are you talking about?" Jason asked, incredulous.

"I'll tell you about it later."

Jane rested for a while in Jason's arms until she felt well enough to continue. Luther and the others did not follow and headed out on their own.

When Jason and Jane entered the nearly abandoned airport terminal, it was pitch black outside, and they were both exhausted. They hunkered down, planning to embark for the summit after they got their breath. They embraced before dozing off from pure exhaustion

.

27

First Summit

Dr. Braxton had implemented elaborate measures to set up this historical worldwide summit. The findings would in theory unify the world communities. He told his staff that he had managed to contact all the world leaders and transport top scientists from all over the world, despite the precarious environmental situation. "Do any of you know of any other organizations that need to be present?" he asked.

"Well, sir, we certainly need to summon the prominent religious leaders throughout the world from every major religion," one of his staffers said.

"I don't think this is a time to get on our knees and pray but a time to take decisive action," Braxton said.

All of his staff frowned and shook their heads. He quickly realized he had better back off.

"Okay, whatever, mates," he said. This was not an act of goodwill on his part but a mandate by those involved in planning the summit.

At the summit, every news station and television network was represented. News reporters were stationed everywhere within the military base. Security was extremely tight, and relative order reigned in anticipation that some meaningful answers might emerge. However, many continued to watch the sky for the ominous sky scrolls and searched for any signs of a quick cloud formation within which sporadic lightning or tornadoes might form.

Outside the complex were countless protesters with signs bearing slogans like "Armageddon!" "Doomsday Is Here!" "Repent! Time Is Running Out!" Others read, "Osama Is Back," Our Ozone Is Depleted," "Global Warming Is Wreaking Havoc," "Scientists Have Screwed Us," and "The Earth Is Breaking Up." The frustrated public was demanding explanations and that something be done. Some were arguing among themselves while others insisted that the government was behind all these occurrences and was trying to cover up some experiment that went wrong. Rioting was imminent, and all-out pandemonium was likely to ensue.

Inside the summit building, Secretary-General Loy peered around the room at the nervous, bewildered, and somber faces. She commenced the program with a polite and cautious thank you to all the countries for coming on such short notice and then outlined the agenda and protocol for the program.

"Let's begin," she said. "I personally want to thank President Clemson for having the foresight and courage to call this meeting. I also want to thank Dr. Norman Braxton for his incredible job in organizing and making all the arrangements for this meeting under a seventy-two-hour gun. I have accepted this mission partly because of my responsibilities as UN secretary-general and partly because of my membership in the human race.

"People in the world are terrified. Some have already experienced tremendous suffering, and I mourn the deaths of their loved ones along with them. This is the most important mission of my life, and I will draw upon all of my wisdom and experience to give assistance to these, the gravest problems ever to face humanity. I have faith that we can work cooperatively to solve these problems. From time to time, I may call upon a higher power rather than any earthly power if required. Let's get right to the issues."

Most of the people in the room rendered polite, conservative applause, hoping for solid answers and resolutions to ease their concerns and fears.

The first presenter was Mr. Thomas James, head of the United States Joint Chiefs of Staff. He began by presenting the information on the screen.

"On March 14 at 11:29 a.m. Pacific Standard Time, a magnitude 4.1 earthquake was recorded on all known seismographs around the world. Also on March 14 at 11:29 a.m. Pacific Standard Time, an exoatmospheric material—i.e., the sky scrolls— appeared as three distinct bands. The opaque, tan-colored bands encircled a 120-degree sector of the earth at 35 degrees latitude spanning from 140 to 30 degrees east longitude—i.e., passing directly through the middle of the United States going eastward to Spain. The scrolls were located approximately 208 miles above the earth.

"Within these sky scroll bands was script similar to handwriting with words that resembled *science* and *God*. Translators now claim these words appeared in English, Japanese, French, and Hebrew. At the same time these scrolls appeared, there was sporadic lightning and a sonic boom over a widespread area that appears to have occurred within the same region as the sky scrolls. The alignment and calibration of most, if not all, geosynchronous earth satellites were continuously drifting eccentrically. Three satellites have now mysteriously disappeared.

"There have been sporadic occurrences of tornadoes in many regions of the world. Often these were abnormal occurrences in the region. Early estimates put the death toll at several thousand. A major oceanic disturbance created an enormous tsunami that originated in the middle of the Pacific Ocean that is traversing slowly, leaving catastrophic annihilation in its wake. At night, a clear shift in the positions of all identifiable star constellations is evident. In addition, a volcanic eruption with massive lava flow has occurred in the middle of the Pacific Ocean along Micronesia on Tavula Island.

"That completes what we know at this time. As additional facts and data are collected and compiled, we will provide them."

The entire room was dead silent as the bewildered members processed the information.

"Thank you, Mr. James," Secretary-General Loy said, breaking the silence. "These are the facts we know thus far. We will now begin the difficult process of attempting to piece together all of these events.

"As you already know, never before have there been so many unexplained devastating events occurring without any rational explanation. Have we seen the worst of these events, or is this just the beginning? We as leaders must show strength and resolve in finding the underlying cause and halting these events. We will proceed in a structured manner, as we would normally do in a crisis. We will allow various entities to come forward and help provide insight. To begin, we will have presentations from selected areas to help assess our situation. Questions and answers during the presentation will follow the standard UN protocol. We will begin with the religious and spiritual perspective provided by Cardinal Robert Malloy, archbishop of the Los

Angeles diocese and renowned world leader. He will provide a perspective developed collectively by several prominent religious leaders."

Secretary-General Loy made it clear that her colleagues selected Cardinal Malloy as the spokesperson representing most of the religious leaders from around the world. "This is a great burden that has been thrust upon him," she continued. "He is challenged to consolidate the views of not only widely disparate religious groups but also gain alignment with the scientists and political leaders in this troubling time. However, with the help of the Almighty, he no doubt will prevail."

Cardinal Malloy stepped to the podium in his brilliant white-and-blue, floor-length robe laced with red and calmly peered around the room. He was tall and thin, and the lines in his face revealed his deep contemplation and worry. In recent years, he had fought with political powers and governments to try to resurrect the respect bestowed on all religious leaders, as the recent religious sex scandals had taken a toll on many prominent religious leaders. Cardinal Malloy was obsessed with his quest of restoring a respectable image, although it compelled open clashes with scientific viewpoints as they continued to encroach the religious realm.

The religious view of God creating the heavens and the earth was becoming more at odds with accepted scientific dogma. Cardinal Malloy was forced to become politically astute and scientifically literate. Further, he adamantly fought with Dr. Braxton and his notion that all physical phenomena and events were explainable as quantitative events capable of accepting measurement. The prevailing public sentiment was there was no need to think that anything or anybody could be causing the mysterious phenomena wreaking havoc everywhere; it had to be a natural phenomenal we didn't yet understand.

"It would be inappropriate and perhaps foolish of me to try to speak with one voice for all the religions of the world. Please pray that the words and information you hear will be beneficial to you and will help guide your understanding. First, I am not here to try to convert or change anyone's views or beliefs. Even some of your top technical leaders have the right to be nonbelievers of faith or hold any religious beliefs."

"Does anyone not know who he is talking about?" one of the representatives whispered to his neighbor. "He might as well come right out and say Dr. Braxton from NASA is an atheist."

"Science encounters difficulty when it comes to explaining some of the most basic questions," the cardinal asserted. "When did the universe begin, what was here before it began, and what was here before that? How will things end? What happens after the end? Why are we here? Are we the only ones here? Should we place the burden solely on science to answer all such questions? Obviously not!"

The cardinal stopped for a moment and surveyed the audience. He walked slowly around the podium and stood before it.

"Most speakers get vibes and cues from an audience. These include eye contact, fidgeting, gazing around the room, nodding, and even sleeping. I only have a limited time to cover some complex, controversial, yet extremely important and profound points. We extracted and condensed them from volumes of books and documents. I sense that it may be extremely difficult to focus in these uncertain times. This uncertainty is causing us to ponder, worry, and focus only on our fears. Therefore, I want to make a plea. Take a deep breath, pay close attention, and listen carefully with an open mind for just ten minutes. I promise it will help."

He stepped back behind the podium.

"Here is the point I want to make. Real belief cannot be supported by evidence from science. Nor can belief be undone by scientific evidence. The sciences deal with what they can see, inspect, and test. But they can make no valid statements about the existence or nonexistence of God because such statements must be made in the absence of any available evidence.

"On the other hand, religion cannot pretend to invalidate the findings of scientists for fear that their belief will be challenged. Math logic says that if the objects of faith are true and the objects of scientific discovery are also true, the objects are equally true and cannot contradict each other. Opposition between religion and science arose from the mistaken notion that religion could present its doctrines as undisputed knowledge that would hold true for all time. The Western religions had incorporated this into its system of belief. For example, this included ancient scientific assertions about the earth and the heavens—i.e., the earth was flat and the center of the solar system, and so on. As these assertions over time proved false, the church reacted because it had mistakenly used ancient science to support its doctrine. In other words, it had attempted to use assumed facts of *science* to support *belief*. It feared, consequently, that if the facts were swept away, belief would crumble. As it happened, religious resistance to science alienated many educated people.

"I believe we are here to carry out God's will and to deliver his message," the cardinal said with a resounding voice.

Several members of the audience sat back and shifted in their seats. Despite his plea, they were not interested in hearing a long, drawn-out religious sermon, but they were willing to endure just about anything if it helped explain the current crisis in the world.

The cardinal looked directly at Dr. Braxton and continued with his delivery.

"The world is experiencing hopelessness and tremendous grief from the grim loss of many loves ones. What is happening to us? Is our fate already sealed? Believe it or not, we have been through this before—several times. To help put things in perspective, we must determine what the various religions of the world have in common. First, they all believe man has some type of spirit and an afterlife, such as salvation, Nirvana, or reincarnation. They all have sacred divine or historical documents, such as the Bible, Qur'an, Vedas, or Torah. Many believe there existed a divine leader on Earth, for example, Christ, Mohammed, or Buddha. And finally, they all believe in some Supreme Being, or beings, such as God, Allah, Yin-Yang, Vishnu, or Shiva.

"Perhaps there is a message written within the sky scrolls analogous to the Ten Commandments," the cardinal continued. "We know it clearly shows two provocative words: *God* and *science*. Is a message actually being sent to all of us?"

"That just about went over my head," a reporter whispered to a colleague as he looked around to see if others were focused on what the cardinal was saying.

"The situation is clear to me, and it is very simple," the cardinal said as he took a drink of water.

"God has shaken us up to get our attention, and he has sent us a message. For the Christian faith, the situation that is confronting mankind today is similar to what Moses faced when he received the Ten Commandments. However, our plight is even more pivotal than what Moses experienced on the mountain or even when he parted the Red Sea. Fortunately, Noah listened to God before the great flood. And fortunately for us all, Noah did not have to apply for a sailing permit or a license to build an ark or hold a summit to get approval to gather the animals. The complex questions we are faced with today have relatively simple answers. Put your belief in God's will first and trust that he will give us answers, help us to make the right decisions, and move us in the right direction. Just believe that God will pull us through all this. That's all it takes. Thank you, and may God be with you."

As most applauded the cardinal's comments, Dr. Braxton stood up in his three-piece suit and gestured to keep things moving to Secretary-General Loy.

28

Face the Reality

As the summit was in session, Jason and Jane arrived, tattered and disheveled and only partially rested after their brief nap. When they saw the crowd surrounding the area, they knew it would be nearly impossible to penetrate the mass of humanity that stood on the brink of pandemonium. Any incident could trigger a chaotic stampede and trample them.

Jason hung onto Jane's hand as they struggled in vain to navigate through the crowd. As they inched their way through, an ominous cloud formation began taking shape overhead. Many people recognized the eerie formation and began to panic. They screamed and scattered for shelter and, in the process, knocked several people to the ground, including Jane, stepping on her back and shoulders. With every ounce of strength he possessed, Jason held onto Jane's hand, managed to get her up, and moved her to a nearby pillar in the middle of the field. She was okay. The panicked crowd continued to disperse to the nearby buildings, leaving many trampled bodies in its wake.

As the crowd thinned, Jason and Jane plowed toward the entrance to the summit. When they encountered a roped-off area monitored by armed guards, they pleaded their case. One of the guards, who was uncomfortable with the growing dissent of the crowd, opened the gate, escorted them to the front door, left them at the entrance, and scampered for cover himself. Again, they were denied entrance, but after identifying themselves, they were allowed into the summit meeting.

Professor Stein sat perspiring at the front table, an ink spot appearing on his shirt pocket. It was clear he would rather have been back at his desk at UCLA solving equations rather than giving a tutorial presentation at the historical summit. He and the others present were unaware of the turmoil outside.

He anxiously scoured the room in hope that Jason, his prize student, would appear. His heart leaped when he saw the disheveled Jason. Professor Stein got the attention of the security officer and pointed out Jason's location. Immediately, the officer guided Jason into the room and to the main table.

Professor Stein took a few minutes to brief Jason on Cardinal Malloy's passionate speech. Next, Professor Stein got the attention of the chairperson, who acknowledged him and informed Secretary-General Loy that Jason Scott from the UCLA research team had arrived with long-awaited news. Secretary-General Loy complimented the cardinal's presentation and indicated they would now learn about a major breakthrough from some members of the scientific community. She introduced Professor Stein, who approached the podium.

"My name is Professor Joseph Stein." He fumbled with the papers in front of him. Also the ink spot on his shirt could be clearly seen. "I am the professor at UCLA directing research related to celestial bodies and their cosmic behavior. The student leading this team is Jason Scott, and he has done some extensive research on the behavior of the earth as a celestial body within our galaxy."

Jason nodded, his heart pounding as he anticipated the reaction to the news he was about to unload on them.

"We appreciate those powerful thoughts from Cardinal Malloy," Professor Stein said as he looked over at him. "However, I think we have some scientific information that needs to be heard before we present any more religious stuff. The findings from Jason's model will explain many of the mysterious phenomena that have brought us together today."

Cardinal Malloy was not sure what the professor meant. His perception was that yet another scientist was berating his beliefs. He was offended by Professor Stein calling his beliefs "stuff" and incensed that no one chastised the professor for such a condescending comment. Annoyed, Cardinal Malloy stood up and walked toward the door. The entire summit body watched him walk away with his eyes trained straight ahead as he abruptly left the chamber, his colorful robe floating behind him.

The absentminded professor was oblivious to Cardinal Malloy's feelings because of his unintentional disrespectful choice of words. "I think we have some important information that will shed significant light on the situation," Professor Stein continued as he looked over at Jason and received a nod of reassurance. "The research we have been conducting at UCLA for the past two years may help explain some of the recent events. I had the research teams from Cal Tech and Oxford verify Jason's findings. We have input the most recent data collected from around the world into our computer model, and it all clearly points to one thing. If you will allow, Jason will explain the findings to you."

Jason came forward and arranged the professor's reference material in front of him.

"First, I want to thank God for allowing me to arrive here safely. Let me give you a quick overview of my research, and this will help you understand the findings.

"My research model incorporates advanced astrophysics and artificial intelligence as the foundations for its algorithms. It revealed that all the data defining the relative position of the earth to the sun, moon, and planets show a consistent shift in position. However, all bodies within the solar system remain in their same relative position within the solar system. What this means is the earth's entire solar orbit has changed significantly. In layman's terms, something has suddenly jolted our planet off its orbit around the sun."

A wave of disbelief rippled throughout the room. The secretary-general pounded the table with a gavel and sounded the alarm bell to restore order. She asked Jason to restate what he had just said.

"Something has moved the earth," Jason continued. "On March 14, at 11:29 a.m. Pacific Standard Time, something jolted our entire planet from its normal position."

"Are you trying to tell us that our entire solar system has been jarred?" the UN Secretary interjected.

"No, I am not saying the entire solar system. Just our planet *within* our solar system. Reputable sources have corroborated my findings. All show that the earth's orbit has been offset by over 1.8 million miles from its normal position. We are showing that this offset will increase to twenty-four million miles as the earth moves around the sun and reaches its aphelion. This is the maximum distance of the earth from the sun."

"What does this mean to us?" the representative from Canada asked.

"We are not sure," Professor Stein said. "We do not know all the implications yet. We are not experts on the meaning of such unexplained phenomena. But one thing we have noticed is that all of our satellite navigational systems have shifted. This is what would occur if the earth shifted, and it is consistent with the facts and data presented earlier. The real impact of this will have to be determined. Undoubtedly, the impact is profound. This shift is the likely cause of the environmental disturbances and chaos. We need a little more time to investigate and evaluate the information and assess the implications, if we can."

"How much time do you need?" Secretary-General Loy inquired.

"I have no idea," Professor Stein said. "If we get complete cooperation from other universities, especially Harvard, as well as the Department of Interior, we should be better able to quantify the orbital deviation in about twenty-four hours. However, to figure out the real consequences of the deviations will take a concerted effort by many others."

"This earth-orbit deviation is one of the most astonishing things I have ever heard," the secretary-general stated. "Please excuse me for a few moments while I confer with the leadership here."

She leaned over and whispered something to those at the head table. They talked for a few moments, and then she came back to the microphone.

"We need to recess this summit temporarily to allow us time to caucus and determine what should be done now. Please remain in the chamber. We will reconvene here within half an hour."

Secretary-General Loy, Dr. Braxton, and a few other leaders quickly adjourned to the adjacent caucus

Face the Reality

room. Jason and Professor Stein were not asked to join them.

In the closed-door session, Dr. Braxton stated his annoyance with the surprise disclosure. He indicated it was impossible for the earth simply to move as the UCLA research squad had claimed. "No one on my technical staff had a chance to review or assess their findings," he growled. "I apologize for subjecting you and the others to such a shocking presentation. In fact, from now on, no one will be allowed to present such claims without having them screened by my experts. The chaos that could ensue from such unfounded and unsubstantiated remarks is enormous."

After Secretary-General Loy challenged Dr. Braxton's harsh assessment, he eventually persuaded her and the other leaders to accept his viewpoint. In truth, before the start of the summit, Secretary-General Loy and Dr. Braxton had met secretly with California earthquake experts to gain a better understanding of the meaning of the worldwide earthquake. In the experts' zeal to explain the significance to the dignitaries, they had speculated the worldwide quake may have destabilized some of California's major faults and that a major earthquake could be imminent. The experts had no idea how much fear and paranoia this created in the minds of the secretary-general and Dr. Braxton. Fueled by this fear, they were willing to use just about any excuse to get out of California and back to the relative stability of New York or Down Under.

Meanwhile, the audience in the chamber room was growing restless, and arguments were breaking out. Finally, Secretary-General Loy, Dr. Braxton, and the other leaders emerged from their caucus and marched back into the chamber.

147

"We have to do better than this," the secretary-general said with irritation. "We cannot allow these bombshell surprises to be presented in this forum. We need to digest the information presented thus far and determine what can and must be done. We do not want to pursue a path of speculation here since the implications of what you are describing are far too profound. We will likely do incalculable damage to the world community if we guess wrongly. We have decided to relocate and reconvene this summit. We will begin arranging to continue the summit at the UN Building in New York since the security and accommodations there are more conducive to a meeting of this caliber. We will ensure safe air travel to all parties participating in the summit. The details of dates, times, and logistics will be posted shortly. At this time, the summit is adjourned."

The membership appeared surprised at the abrupt ending yet agreed that it was the right decision. But Jason and Professor Stein were stunned.

"I'll keep my damned mouth shut next time," Jason whispered to Professor Stein. "They just implied our work is nothing but speculation. Now they're going to send everybody across the country in these dangerous conditions. Are they threatened by reality? We ought to back off and stay here."

29

Rescue Readiness

The entire world was tuned in, waiting for any breaking news from the world summit at Vandenberg. Among them was Captain Norman Atkins, a navy rescue helicopter pilot who sat with his team members in a staging building on a military base near the Hawaiian Islands. They were on standby, awaiting orders from Dr. Strickland of NOAA. They were poised to embark upon a rescue mission to transport as many civilians as possible from the island to a naval fleet in the area.

The strong faith required for the mission was not a characteristic held by Captain Atkins, a tough veteran rescue pilot with a proven record. The previous year, he had lost his wife and daughter in an accident that left him so embittered that he had renounced his faith. His recently developed atheistic views aligned him with those of Dr. Braxton. They both championed the notion that they needed no help from a higher power; sheer determination was adequate for all endeavors.

Atkins and his team expected dispatch instructions soon to the islands near the slowly encroaching ocean swell. They sat with their eyes glued to the TV.

Cardinal Malloy addressed the worldwide television audience about a vigil, which was to commence from his cathedral in Anaheim as soon as he arrived from the summit. An assemblage of world religious leaders awaited him, preparing to start the special televised nightly vigil that was to continue until they received a message from the UN authorities. They wanted all nations and scientists to cease all activities known to cause serious ecologic damage and pollution to God's planet. The pope, the Dalai Lama, and most of the major religious leaders from around the world were on standby in their respective countries as well. They had been caucusing and praying for an abrupt halt to ecological destruction and for peace throughout the world. A contingent of people from all walks of life had gathered inside and outside of the cardinal's cathedral. Further, people of all religious faiths had gathered at various sites on all continents.

Cardinal Edward M. Egan, archbishop of the New York diocese, kicked off the vigil. He was an expert on many so-called scientific principles and was himself a renowned scientist. He was also vocal on his disapproval of Dr. Braxton's condescending views on religion. He believed a person with overt atheistic views should not hold such a prominent leadership position.

"All spiritual leaders are in agreement with the message you are about to hear," Cardinal Egan declared from his broadcast podium in the cathedral. "God has asked me to deliver this message to you, and each of you must take action."

"I don't want to hear another religious speech," Captain Atkins interjected. He started to change the channel, but one of his team members insisted that he wait so they could hear what the cardinal had to say. The captain reluctantly backed down. The same broadcast was on nearly every channel anyway.

"Our scientists are doing their best," Cardinal Egan said. "They want to convince us that something has moved our entire planet from its normal orbit around the sun. However, even within the scientific community this is widely considered impossible based on what we know and understand. Those of us familiar with basic astronomy have been able to see for ourselves that something has happened. The stars at night look different, and scientists admit there is no plausible scientific explanation. Yet there is abundant evidence confirming that something serious has occurred. Our own eyes tell us something is amiss. Some bizarre explanations coming from the scientists and the news media include, one, our planet, Earth, is breaking apart; two, our planet is entering a new ice age; and three, alien forces are pulling our entire planet away from the sun.

"We believe this alien force theory is not new to some of us. Could this mysterious force be our Maker sending us a message? Scientists will always seek to explain events they don't understand with hypothetical scientific explanations. They first make a hypothesis, they then do some experiments and examine the results, and if the results agree with the experiment, it becomes a theory and then a law, such as Newton's laws, Kepler's laws, and so forth.

"But this scientific method can only carry us so far. In looking at the origin of the universe, scientists have concocted the Big Bang theory. But they cannot explain what existed before this big bang. They also cannot explain how the universe will end and what the end will even mean. Many of them don't realize it is okay to say "I don't know" to some of life's questions. Some scientists are still looking for boat fragments to verify that Noah's ark in fact existed. Others study wind conditions and tidal wave behavior to explain how Moses parted the Red Sea. Some even search for a scientific explanation of how mysterious writing could have suddenly appeared on the stone tablets of the Ten Commandments. They will never agree about divine evidence and will always belittle these divine events. Every parable in the Bible, Koran, Vedas, and the like will likely remain under investigation forever by many of our so-called astute scientists."

Captain Atkins crossed his arms and rolled his eyes.

Cardinal Egan continued. "The only issues today in which no know scientific experiments are underway are those involving the existence of God. However, many scientists accept the reality of death but question the possibility of an afterlife, an after-place, such as heaven, hell, purgatory, and the existence of a spirit. Regrettably, many scientists do not accept the existence of miracles and claim scientific reasoning will eventually explain away all miracles. As you know, in the Catholic faith we grant sainthood to those who have performed three miracles. However, many scientists discredit saints, claiming the miracles ascribed to them arose from luck or a fortunate event."

Captain Atkins stared at the red phone, willing it to ring so he could stop listening to the cardinal's diatribe.

"People of today are veracious consumers of the world's resources," the cardinal went on. "Natural resources show depletion at alarming rates. Government and corporate scientists are working feverishly to satisfy these appetites. We can meet the needs now and hope for future resurrection of depleted resources. But if man cannot find a way, we will need to rely on miracles."

Captain Atkins looked around at his team. "This guy is really relying on miracles.

"We may not be able to resolve the differences between science and religion in the near term, or perhaps ever," the cardinal said. "However, we must continue to try. We will continue our vigil by keeping our doors open all night and will remain here until we receive a message from the UN that our world leaders will cease destroying our planet. This may mean ceasing space flight, even all aircraft flight, halting destruction of the rain forest, or banning pollution from vehicle exhaust.

"Despite our differences, we can still love one another and work together to solve our problems. We ask all nations to join us and pray that our well-intended efforts are not divisive. Only time will tell."

Captain Atkins stared down his team members. "Why did you guys waste my time with this vigil stuff? I didn't want to listen to that crap. I have never heard so much bull. A vigil to wait for the UN to stop scientists from doing research, saints and miracles—come on, give me a break! We save lives based on our knowledge and ability. We can't start hoping for a miracle!"

"The captain is way too bitter now," one of the team members mumbled to his friend. "His faith is gone. The loss of his family has taken a toll on him."

"You're right. I hope he doesn't jump the gun on the rescue mission. It looks like a miracle is exactly what we'll need to pull this off."

30

Ocean Swell Devastation

Hundreds had bravely and nervously gathered outside the National Oceanic and Atmospheric Administration building in Bethesda as the summit was underway.

"We can't hold off any longer," Dr. Strickland said as he peered down at the table at his staff. They eyed him anxiously, wondering if his delays and hesitancy to take aggressive action would cost lives.

"We must issue a full-scale alert," he stated. "The threat is real even though we have no clue what caused these catastrophic conditions. We must act now and pray we don't send our men into harm's way. Weather bureaus from our sister networks are beginning to report on the massive ocean disturbance."

Dr. Strickland took a deep breath before walking outside the building to give the press a forthright account. He took his position in front of the array of microphones.

"The president of the United States is at the emergency summit and was just briefed on the impending devastation. I have alerted the emergency rescue resources, including the helicopter rescue squadron. They are on standby."

"Ladies and gentlemen, we are issuing a category-five emergency weather alert. We have detected a tidal wave approximately fifteen hundred feet high and over two thousand miles long moving in a westerly direction at approximately twenty knots. We believe this constitutes a mega-tsunami of a size never before recorded. Expected landfall on the western portion of the Pacific Rim countries we estimate to occur in two days and it is likely to engulf many of the eastern islands including Korea, Japan, and the Philippines. We predict many countries on the western side of the Pacific Rim, including the United States and Canada, will not experience the tidal waves but will experience enormous low tides that will wreak havoc on our shores and docks.

"It appears that some as-yet-unknown phenomena has jolted the entire Pacific Ocean to slosh, resulting in this immense swell propagating the entire length of the ocean. All countries on the west side of the Pacific Rim have received notification to evacuate all coastal shore cities. Deployment is underway for rescue helicopters and other aircraft to help evacuate as many people as possible from all the affected islands. We don't know what they might face, but we must try to help our citizens of Hawaii the best way we can. We are deploying every local and national governmental resource to assist. Are there any questions?"

Uproar went up from the news reporters, and for a time, order was lost.

Hurricane and other weather alert warnings, the National Guard, all military personnel, and the UN civil defense system initiated activation. Immediate evacuation and rescue of all personnel from the mid-Pacific Ocean region began. All the Pacific Rim countries activated their civil defense systems and prepared for the worst.

Sheer panic and confusion spread throughout the world.

Aphelion

31

Rescue Efforts

As the squadrons of American-led rescue helicopters approached the Hawaiian Islands, they witnessed the enormous, slow-moving ocean swell that threatened to submerge the entire chain of islands within hours. Any attempt to rescue a significant number of people would most likely be futile. Nevertheless, in spite of the dangerous sporadic tornadoes and unpredictable lightning strikes, Captain Atkins ordered his entire fleet to the opposite end of the chain to attempt a one-time rescue of as many people as possible.

As the copters swooped down on the far island of Honolulu, many of the terrified residents began to scatter. Either they had not been warned or they failed to believe the impending disaster about to befall them.

When Captain Atkins landed, he heard the emergency response sirens in the background. Despite this, the people did not know what to do. He frantically shouted over his loudspeaker for people to drop everything they were doing and get into the helicopters. Several residents backed away from the helicopter, reluctant to approach. Annoyed that they didn't realize he was there to help, Captain Atkins ran toward the frightened residents, who scattered into the foliage and dense, towering trees.

As he tried to grab some of the children, he heard a thunderous roar in the distance. He let the kids go and moved out from under the trees. As he continued in vain to convince the people to come onboard, he looked over his left shoulder and gasped.

A mountain of water towering hundreds of feet into the air was approaching with ominous power. He quickly determined that he might not have time to scramble back to the safety of the helicopter before the water descended. And if he did make it, the helicopter might not be able to become airborne in time.

Captain Atkins dropped his loudspeaker and ran toward the helicopter. As he leaped over some shrubbery, he saw two toddler girls huddled together among some bushes. He stopped, picked them up, and resumed his dash to the helicopter. As he did so, he heard the trees just behind him starting to crack and topple.

His right foot clipped the top of some bushes, and he and the girls tumbled to the ground in a ball of flailing arms and legs with the girls flying in opposite directions. One of the girls hit her head on a rock and lay motionless while the other girl popped up and began to cry. The water mountain was so close he could feel the spray on the back of his neck. Instinctively, Captain Atkins snatched the crying girl by the arm, and with every ounce of strength he had, coupled with stark fear, he dashed to the helicopter.

"Get the hell out of here now!" the captain shouted to the pilot, who was oblivious to their imminent annihilation.

As the startled pilot attempted to maneuver away, they heard the roaring wall of water just behind them, demolishing everything in its path. With the door still open, the helicopter jolted and became airborne, jerking the little girl away from Captain Atkins. She grabbed the helicopter's landing rail and hung on for dear life, screaming hysterically. Captain Atkins froze, looking into her eyes. He then desperately lunged to grab her, but before he could reach her, the turbulence blew her off the rail. He saw the horror in her eyes as she flew into the abyss.

The helicopter strained to outrun the oncoming ocean water. The co-captain radioed to the other helicopters in the fleet. "Mayday! Mayday! Leave the area immediately!"

Captain Atkins watched in horror as the mountain of water swallowed most of his fleet. As his craft sped away from the island, he thought of his wife and daughter who had been killed in a freak drowning accident the year before. Motionless, he looked on in shock as the crushing water mass overtook the last of the Hawaiian Islands.

Not only had he failed to save a single person, he had lost most of his squadron. Captain Atkins felt the grave irretrievability of the loss to the depths of his soul.

32

The President Directs Experiment

President Clemson remained at a wrap-up session at the adjourned summit. She was under tremendous pressure to determine the cause of the global tragedies and if more disasters were imminent. After being escorted into a special security office building on the Vandenberg Air Force Base, she caucused with her advisory council and brainstormed various options.

Dr. Braxton and the head of the Joint Chiefs of Staff, Thomas James, emerged from another sidebar meeting following the summit.

"NASA's advisory council recommends a special experiment be conducted," James said to President Clemson. "It will help us understand the calamities and possibly forecast what might lie ahead for the planet."

"These puzzling phenomena must be found and communicated to the world, with the assurance nothing else is imminent," the president said.

"There're all related," James said. "We now know the earthquakes, sky scrolls, shifting stars, and ocean swell all occurred at the same time. It can't be a coincidence. Clearly, something happened, or is happening, to this planet. The mega-tsunami currently wreaking havoc on Hawaii is also likely related."

"The chap is bloody right," said Braxton. "Madam President, NASA is recommending we conduct a unique experiment to obtain some vital data that may explain our situation. We need to get a better look at our planet from a specific distance and take some measurements. We can do this using the equipment onboard the space station currently orbiting the earth. However, we need to place the station into a highly elliptical orbit perpendicular to the plane of our solar system. In other words, send it far enough away from Earth so it can obtain some crucial data uncontaminated by our atmosphere."

"What exactly do you expect to accomplish with this experiment?" President Clemson asked.

"We believe the space station can detect any changes in the shape of the earth and the position of the planet relative to other celestial bodies. We believe some magnetic or gravitational field from the other planets may be affecting us, or perhaps the sun is undergoing a change and causing the earth to shift. More importantly, we need to look for any movement in the earth's tectonic plates. This experiment will allow us to get some measurements that are not influenced or polluted by Earth's environment or ecosystem. And we need to proceed immediately."

The president's chief financial officer, Randy Fillmore, shook his head. He was a short, thin scrawny man. He was often uneasy about new ideas without a thorough investigation. He was said to be both a smart-ass and a kiss-ass. For him, every decision revolved around money. Officer Fillmore had recently been at odds with Braxton, accusing him of spending money on any harebrained idea without regard for fiscal responsibility. Fillmore thought he was cabinet-level material, but he often stepped over the line relative to his authority with the president. He stared at Braxton from across the table.

"Excuse me, Mr. Director, but as I recall, the space station lacks the ability to move around at the whim of some new command."

"No, it now has the ability to relocate," Braxton said. "I am not sure where you have been for the last few years."

"I have been right here, doing my job."

"Then aren't you aware of the EMRS project?"

"Mr. Director, I have approved funding for your little project. However, the sparse technical details have always been cloudy at best. Would it be asking too much to enlighten us again, sir?"

"It stands for Enhanced Maneuvering Retrofit System."

"I know that, Mr. Director. Can you briefly describe its purpose and claim to fame?"

"Yes, sir. It is designed to give the space station the ability to move into a different orbit around the earth."

"Okay, I know a little about that part. So what does that mean?"

"It means the space station now has the ability to move into a geosynchronous orbit, an HEO or LEO orbit, depending on the requirements of the mission."

"Well, that's good to know. So what good is that?"

"Putting the ISS into geosyn will allow us to take Earth's measurements to the accuracy needed, as I just stated. Perhaps you were not listening."

"Excuse me, Director Braxton, but I would think at this point changing plans on a whim might be a tad risky and overzealous. You know we have a few people in there."

"The only minor risk or concern is that the autopilot system on the ISS requires a great deal of structural stability."

"That sounds like an unacceptable risk to me."

The President Directs Experiment

"No, it's very minor. The potential for its becoming unstable has always existed. However, as a precaution, most of the onboard systems become disabled until the maneuver is complete."

"What does all this mean to the ISS crew members?" Officer Fillmore asked.

"Wait a minute, am I on trial?" Braxton asked. "Look, I am concerned about the safety of the ISS and its crew. Significant disturbances, such as any flexing of the solar panels, may create ISS maneuvering stability difficulties. This means that too many crew members moving at the same time may cause problems if they do so while the maneuver is underway. Therefore, it disables most of the systems onboard during the maneuver. Nonetheless, prior testing shows that the system works fine. Let's move on!"

"I have heard enough," Officer Fillmore said. "There is no way we're going to divert the space station into a radically different orbit by sacrificing our entire space program to conduct some far-fetched experiment within such a short time frame. Besides, our space station plans have taken years to evolve, and we don't have the funds for this sort of thing. So scrap the idea."

President Clemson could no longer contain her exasperation. Suddenly losing all dignity, she lunged across the table toward Fillmore. One of her shoes flew off, and she tore her stockings and scraped her knee. Fillmore fell back into a chair as the president got one hand around his neck and grasped his shirt collar with the other.

"You stupid jerk!" she shouted in his face. "You had better figure out a way to help get this done and stop being a hindrance! I can't believe you are focused on funding. We have a world crisis that is affecting all of mankind. The entire planet may be about to explode or collapse. People are panicking, entire islands are destroyed, the world is experiencing bedlam of unprecedented proportions—and you're worried about a stupid budget! We cannot continue to sit around and talk about what we cannot do. You had better get with the program and help figure out a plan!"

The president released Fillmore. She was traumatized by the deadly tsunami, the lightning and tornadoes, the sinister sky scrolls, and the distressing uncertainty and mounting fear of the future. Was God punishing mankind? Should she heed what was said at Cardinal Malloy's vigil—to focus on ceasing all environmental activities that might be detrimental to the environment and ecosystem?

The president froze and looked around the room. Everyone looked bewildered and anxious. Feeling pressured to provide quick guidance, her anger soon gave way to a sense of helplessness. This weighed heavily on her, so much so that her legs buckled and she dropped to both knees. She covered her face with her hands in an attempt to conceal her distress as tears began streaming down her face. Several of the men in the room also began to succumb to feelings of helplessness and the situational pressure, but rather than comfort the president, they slumped with their heads down on the table. After a few moments, however, the president found the inner strength and composed herself.

"We must pull ourselves together and go forward. If this experiment is the right thing to do, let's do it. We have to do something, and we have to do it now!"

The President Directs Experiment

In the back of his mind, Braxton weighed the distress against the urge to shout with delight after witnessing the president's attack on his nemesis, Fillmore. But he concealed his feelings and gathered his team, including the shaken financial officer, and shuttled down the hall into a room where they commenced to brainstorm about what to do.

"First, does anyone know any technical reason why we can't conduct this experiment? I understand we have enough fuel onboard the space station to achieve a high enough orbit. I am aware, however, of ground station's concern that we need to closely monitor calibration issues associated with the maneuver. Other than that, does anybody know of any showstoppers? I can't see us conceding to the demands from the vigil and curtailing all space experiments at this point in time."

Everyone agreed, as they knew of no reason to not go forward. Braxton and his advisory members proceeded to evolve the details of the plan.

Upon consensus, they used the closed-circuit channel to get in direct contact with the crew of the space station and explain the situation. Braxton then asked the crew members some direct questions.

"Do any of you feel reluctant to conduct the experiment we've just proposed?"

A somber mood had prevailed onboard the space station since the dreadful docking incident and Larry's demise. Nonetheless, they discussed the ISS maneuver with Braxton and his staff and agreed to brief the president on their recommendations. They set up a multiparty communication among Dr. Braxton, his staff, and the president at Vandenberg; NASA mission control in Houston; and the crew members of the International Space Station.

"We believe this mission is vital to the world," the ISS captain announced. "It has also been pointed out to us that if serious events continue to escalate back on Earth, we may be the only humans poised to survive. The destruction of the Hawaiian Islands and the impending devastation of many of the Pacific Rim countries make taking a risk with the ISS almost meaningless."

NASA's chief scientist said, "I agree. I think we should be able to put the ISS into this altered elliptical orbit in a timely manner."

"I'm sure you understand we are still recovering from Larry's death," Dorothy the ISS captain said. "This is very important to us up here. I believe you know we have talked to Larry's wife, Bennie. She has remained remarkably calm; however, we understand that she has gone into seclusion until we return with Larry's body."

"Yes, we are aware," Braxton said. "Bennie has expressed a strong desire to obtain the caricatures that I sketched of Larry. You know, the one depicting him riding the space motorcycle with his arms raised as if on a bucking bronco."

"I am aware of this request," Dorothy said. "I will be honored to give the caricature to Bennie upon our return. Despite all of our tragedies, our team has discussed this maneuver. We believe we must do this for Larry and the rest of mankind. We accept the risk and stand by, awaiting directions from you to proceed."

After signing off, Braxton led his team and Officer Fillmore back to where President Clemson was waiting for his decision.

"Okay, Madam President," Braxton said, looking around the room. "I recommend we proceed with the experiment."

The President Directs Experiment

The president was surprised by Braxton's quick decision. She felt she should not waiver by suggesting further evaluations, so she simply nodded in approval.

Preparations immediately ensued, and the countdown for the maneuver began.

33

Space Station Experiment Commences

The morning following the summit, only a few people of many of the coastal cities were taking precautionary measures of leaving their homes and moving inland to hotels, evacuation centers, or staying with friends. Others were waiting for some clear direction or instruction.

Jason, Jane, and Professor Stein went back to the UCLA lab where the research team was busy collecting data and running computer simulations from the celestial body movement program. As they reviewed the results, the existence of a new earth orbit around the sun was confirmed. Jason asked the lab secretary to summon his colleagues.

"I have the results," Jason said as they all arrived, including Austin. "They show the earth and moon are now moving in a shifted elliptical orbit around the sun. Look at the size of the earth's aphelion—the farthest distance from sun." All heads turned to his video simulation on the computer monitor. "It is normally 94.5 million miles from the sun, but now we can see it is projected to be close to 120 million miles. The perihelion—the closest distance to the sun—is normally 91.4 million miles, and now it is projected to be only eighty-five million miles."

"What's happenin'? That's hard to believe," Austin said. "What are the implications of these orbital changes, dude?"

"Not sure," Jason replied. "The only thing I can say is this is exactly why all the earth satellite systems are screwed up."

Another student chimed in. "What do you mean? What do you think will happen to all of our earth satellite systems?"

"Who really knows? I'm not sure, but since they look at star positions relative to the earth's position, the satellites will likely try to self-correct their positions. But they may not be able to achieve this. In other words, the satellites will look at Earth and then look at a star. They will detect that things are different from what they are supposed to be and will tell the autopilot system to keep changing until its measurements are corrected. They will keep trying, in vain, to correct when they can't. Furthermore, any new systems undergoing maneuvers may not be able to self-correct."

"That's because the new systems use the AutoNav autopilot system," Austin pointed out. "I'll bet our research model is the only informational system in the country that uses artificial intelligence to correctly predict satellite motion due to such an unfamiliar situation."

"What Austin just said is very important," Jason asserted. "None of the government or NASA models could predict this behavior. I hope you realize this is a key by-product of this research."

It was Professor Stein's turn. "If this AutoNav problem is true, I'm concerned with the space station experiment."

Puzzled, Jason looked at Professor Stein and paused before asking in a stern voice, "What space station experiment?"

Professor Stein suddenly realized that Jason had not heard about the decision to enact the experiment that involved his parents. "The ISS experiment," he said. "Last night, the president and NASA staff decided to put the space station into an orbit considerably out of the plane of our solar system and perpendicular to it."

"What??" Jason exclaimed. "What are you talking about? They can't do that! That would put the entire space station and my parents in serious danger!"

"That's true," Professor Stein said.

"Damn it!" Jason slammed his fist on the table. "You have got to be kidding! They would *not* do something as dumb as that!"

No one in the lab said a word.

"Stupid fools!" Jason shouted. "We have to stop them. The entire space station could fly off course and never return!" He slammed his book on the table. "Professor Stein, you must contact NASA and tell them the dangers."

Jason and Professor Stein rushed to the phone to try to get a message through to Tom James. They finally got through to him only to learn that the automatic countdown for the mission maneuver had already begun. The only way to stop the experiment was to contact mission control directly. However, despite intervention from Mr. James, the countdown completed, and the automatic sequencing maneuvers for the space station began.

Jason and the others in the lab were able to intercept the video signal as the entire NASA community sat glued to the TV watching the space station undergo the maneuver placing it into an elliptical orbit. Jason's mother, Dorothy, would assist in taking spectrographic measurements to determine the content of the sky scrolls. Although his father, Walter, had not fully recovered emotionally from

Larry's death, he was well enough to contribute to their efforts.

At mission control, a NASA official made some disparaging comments about Walter earlier in the day. He implied that Walter's reaction to Larry's death showed a breach in professional demeanor that indicated a serious psychological compromise.

"No, I disagree," another NASA official said. "I think Walter's reactions reveal a normal human response to a very tragic event. How he recovers from the trauma is more important."

After sidebar consultations with the NASA psychiatrist, they concurred that not only would Walter's participation in the earth measurement experiment be vital it would also be therapeutic for him.

Several others of the six-member crew were also involved in the main experiment of making observations and taking measurements of the earth's exact size and shape over a short period of time. Crucial to the heart of the experiment was determining if the earth were undergoing any unusual pulsation or tectonic-plate movement.

"The experiment will answer some puzzling questions," one NASA scientist promised. "We now know the liquid core of the earth moves and sloshes. This causes a gradual change in the earth's magnetic field. We have been studying this magnetic field for many years now and know it can change and has in fact reversed many times during its existence. In fact, it undergoes a complete reversal every several hundred thousand years. The last one occurred about three hundred thousand years ago. This means a common magnet pointing in one direction will point in the opposite direction—north will become south. Preliminary evidence indicates our planet is in the midst of another magnetic-pole reversal."

"And further," another scientist said, "if they determine that one of the earth's seven plates is undergoing erratic behavior, perhaps they can relocate people to avert earthquake fatalities. We think these and other measurements will help get to the bottom of the disturbance phenomena plaguing the planet."

As the community watched, the space station began to power farther and farther away from Earth. Suddenly, there was a burst of static over NASA's closed-circuit TV and an abrupt interruption in the signal. For several minutes, every tracking, relay satellite, and ground-tracking system scrambled to locate the space station and determine why communication had been interrupted. As everyone anxiously waited for NASA to restore communication, mission control finally announced to the public that coverage was suspended for the time being due to technical difficulties.

Jason and his research team, however, had a direct link to NASA's secure communication line and heard the real news: Mission control announced that communication with the space station had been lost. It appeared that it ejected away from Earth and was headed away from our solar system. The AutoNav was trying to correct for uncorrectable misalignments and consequently had shut down onboard communication until it believed the alignment of the space station maneuver had been corrected.

The NASA spokesperson and much of the scientific community were baffled. The NASA communication briefing at mission control erupted in turmoil. It was clear to Jason and the UCLA researchers that the AutoNav was working against

NASA when it came to the safety of the space station and the astronauts onboard.

Jason was stunned and devastated. "Oh my God! Mommy! Daddy! Why are they conducting this insane experiment?"

NASA's ground station reported that the space station was streaking away from Earth's solar system, although it did not appear damaged. After a few hours of attempting to reestablish communication, it appeared that all hope was lost.

Jason finally mustered the strength to call Britney to tell her the bad news. Immediately, she became hysterical and blamed NASA.

"Those stupid jerks!" she screamed over the phone. "I had confidence in them like Daddy said. This is all because we didn't agree with their stupid policy of not allowing both parents on the same mission. I never dreamed they would pull this kind of stunt! Those bastards have gone too far!"

In the days that followed, Britney withdrew from her friends. The impact of the tragedy gripped her mind, which resurrected every leisure moment. Consumed by grief, she had no will to get better or even to go on living.

34

Brainstorming The Mystery

Jason and Jane were overcome with grief at the loss of his parents. They and some of Jason's colleagues spent Saturday night in the lab in mourning and were moping about early Sunday morning, barely speaking to one another. They could see periodic flashes of lightning through the window and were in constant fear of a surprise tornado ripping through their lab.

In order to continue to understand what was happening in space they had been in constant contact with NASA. It just did not seem possible that it was swept out of the earth's solar system. With the AutoNav stuck in a controlled, self-correction mode, it believed there was a misalignment between the earth and the stars it used to align itself. In vain, it was trying to correct the uncorrectable.

Two of the colleagues began whispering to each other. "I didn't sleep well last night, but I don't mind. I would rather be sleeping here on the floor with the team. I'm glad all of us are together."

"I just hope NASA fesses up and confirms what happened with the space station."

Another student approached and sat down on the floor with them. "I have a question. If the space station isn't damaged, how long could Jason's parents and the crew last?"

"In theory, the space station could last a long time," one of the students replied. "But I don't know. Micrometeorites and other external effects will likely deteriorate the structure in a few years. Their food and other supplies can be recycled and can last many months, maybe even years."

"I pray something miraculous happens. I wonder what it's like on the space station now?"

"Jason's father, Walter, once described it to me. He said it's like taking a flight on a 747, except the flight lasts six months. He said the combined crew compartments are like maneuvering around in two 747 passenger sections tied together."

Chairman Chapman entered the lab and joined Jason, Jane, Austin, Professor Stein, and the other lab members, chewing gum to mask the alcohol on his breath. The students continued to quietly mingle and cogitate in the lab throughout the morning.

"What is happening?" Jane blurted out, finally breaking the tense silence. "Does anyone have any idea why all of these things are happening to our planet?"

No one answered.

"Do you know what Cardinal Malloy said?" Jane continued. "He, the pope, and the world religious leaders all believe God is giving us a wake-up call. I hear these same comments echoing throughout the entire religious community. I think people of all faiths believe that this is God's will. No one can explain what is happening based on any scientific laws."

"That is bull crap, Jane!" Jason shouted. "There has to be a scientific explanation for this. Damn it! I don't know what it is, but there has to be some explanation that I can understand. Neither God nor anybody else is explaining what is happening to us. We need to determine our own fate and answer things for ourselves."

Everyone grew silent, sympathetic to Jason's uncharacteristic outburst. They kept quiet, hoping he would simmer down.

The research team had special access to much of the raw data exchanged among the universities and government research labs, such as the Lawrence Livermore National Laboratory and NASA's Johnson Space Center. They also had direct access to the raw data transmitted from the Hubble telescope orbiting the earth. In fact, one of the research students had been intercepting some of the Hubble data.

"I have noticed something strange," the student said. "Hubble has recorded the presence of a massive object having passed near our solar system a few days ago. The telescope took a bizarre picture that revealed an unknown object blocking a large portion of the sky."

"Do you think it has anything to do with what has been happening to us?" Jane asked.

"I'm not sure," he replied. "But I do know that several astronomers have been reporting similar unexplained observations. Just before the devastation in Hawaii, the Keck optical telescope had also revealed a huge object blocking the view of many stars. This only lasted a fraction of a second, but it occurred at the exact instant as the Hubble sighting. And both sightings occurred at the same time as the earthquake. It appears that something invisible and enormous was there and moving past the earth at an incredible speed."

"That's it!" Jason shouted, jumping up. He paused for a moment, thinking. "I think I know what the telescopes were seeing! A rogue black-hole fragment! A black-hole fragment or a particle from a blazar-type quasar streaking by our solar system nearly hit us. That's it!"

Brainstorming The Mystery

Everyone in the lab stared at him, intrigued.

"It all fits." Jason paced the room as his thoughts began to coalesce. "We have suspected that plasma jet streams from blazars, gamma ray bursts, and fragments may be streaking through our galaxy toward us. One of these fragments may have glanced by our planet. It probably acted like a comet and knocked our entire planet out of position, bumping us out of our orbit, and was probably so massive that its gravitation dragged the earth with it for a while. Just like what happens when you brush a magnet by a compass—it spins."

"That's incredible!" Austin blurted out. "I think you are on to something, Jason. Brilliant!"

"It's all starting to make sense," Jason said, his eyes shining with excitement. "I believe this explains some, if not all, of these phenomena we have been experiencing. The jolting of the entire earth probably caused the earthquakes and the ocean swell."

"And the lightning and tornadoes," another student said. "They were caused by the outer atmosphere being churned up from a near miss."

"Things are falling into place," Jason said. "If something glanced by our planet, it is likely all of these things would have occurred."

"Hold on a moment," Chairman Chapman interjected. "That is preposterous, Jason! I have never heard such far-fetched notions in my entire life."

Everyone stopped in their tracks and looked at the chairman.

"You guys are grasping at straws," he continued. "You are not thinking clearly. Jason, I think the loss of your parents is clouding your thinking."

"But, sir," Austin said. "I think Jason's theory has serious merit based on what I know about black holes and blazar jet emissions."

"Come on now, Austin," Chairman Chapman scolded. "How can you think that? You guys have to get a hold of yourselves and take a break."

"Sir, this is not preposterous," Austin said. "I for one am not grasping at straws. This is what I have been focusing my research on for years now. I know what this stuff is all about."

"Think about what you're saying," Chairman Chapman asserted. "Do you folks know what the hell you are talking about?"

"Yeah, we damned sure do," Austin said. "Let me tell you what I have learned. A blazar is a very compact quasar with a super massive black hole at its center. It emits a relativistic jet pointing in the general direction of the earth. A black hole's mass has begun to accrete into a finite region. Within a certain distance from the center of the black hole, the gravitational pull is so great that nothing, not even light, can escape. That distance is the event horizon. The event horizon is not a physical boundary but the point of no return for anything that comes inside it. When scientists talk about the size of a black hole, they are referring to the size of the event horizon. The more mass contained within the center the larger the event horizon."

"I am familiar with all of that," the chairman said. "So what is the connection?"

"The connection is simple," Austin said as he paced the floor. "I have been studying the black hole designated as GRS 1915+105, located forty thousand light-years away. It first came onto the astronomical scene in 1992 with a burst of X-rays believed to be caused by the black hole suddenly swallowing hot gases. The black hole in GRS 1915+105 is purported to comprise ten times the mass of our sun compressed into a twenty-five-mile-wide sphere. That is incredible! Since 1992, astronomers have also observed the system fire off a couple of gigantic jets of hot gas and fragments from the dense core into outer space. It appears to me, like Jason said, some of this black-hole debris has been heading our way for some time now and has finally executed a near miss, in galactic terms, of our planet. We all know that only X-rays are believed to be ejected from black holes, but who knows what quantum interactions with mass actually occured."

"Aha!" Jason said. "This helps explain the occurrence of the sky scrolls. A fragment of black-hole debris must have whizzed by our planet. It scorched our atmosphere, causing the sky scrolls, while the strong magnetic fields of the fragment dragged our entire planet out of position."

"You're right!" Austin said. "This black-hole fragment was apparently moving near the speed of light so our telescopes never saw it coming even though it probably has been heading our way for thousands of years. When it whizzed past the earth, it jarred us, creating the jolting sensation we characterized as the earthquake felt around the world. The intense gravitation pulled the Pacific Ocean, just as the moon does, and created the massive ocean swells."

"So, Mr. Experts," Chairman Chapman said, "you have all the answers. Do you really believe all this bullshit? I have never heard such double-talk before. Do you think any sane person would understand, let alone believe, your rambling high-tech explanations?"

"But, sir," another student interjected. "It all makes sense to me too."

"I must agree," asserted Professor Stein, joining the ranks of black-hole-fragment believers. "It initially sounded bizarre to me, too, but based on all the facts we are aware of and the science that exists today, I believe this is a plausible explanation. We scientists must keep an open mind to things we have not yet experienced."

"You're all going off the deep end!" Chairman Chapman angrily exclaimed. "Okay, enough is enough. All of you have been working on this project too long, and your thinking is distorted. In my opinion, you are on the verge of a mental breakdown."

"Shut the hell up!" yelled Professor Stein. "I think your sorry ass had better face reality. For one thing, something way out of the ordinary has occurred on our planet. Whatever it is defies all scientific explanation. There is no such thing as a worldwide earthquake. It is impossible, yet it occurred. It is not possible for the entire Pacific Ocean to suddenly split apart and slosh over nearly its entire length, yet that occurred. Tornadoes and lighting don't just come out of nowhere. These extraordinary occurrences were created by an extraordinary event. It may be due a black hole fragment or even some bizarre clump of invisible plasma dark matter. We just don't know. But I think we have a viable answer. You had better take a mental break yourself and trot your ass back to your office where you can stew in your conservative, old-fashioned thinking and hit the Jack Daniel's."

Stunned, Chairman Chapman glared at the professor and stormed out of the lab. Looking back over his shoulder, he said, "It doesn't matter what I think. I am certain no one else will ever believe your crock of hysterical claims. I would be embarrassed to tell anyone what you have just told me."

A few hours later, Professor Stein decided to visit the chairman. In doing so, he had to brave the treacherous environment en route to the chairman's office. Using his cane, he quickly trotted toward the engineering building, Bolter Hall, frequently scanning the skies for the sky scrolls and watching for any sudden cloud formations or sporadic lightning. He noted the large tree, Brutus that had fallen against the classroom during the earthquake.

The professor's purpose for visiting the chairman was to make amends for his caustic comments regarding the revelation of the streaking black-hole mass.

"How are you holding up now?" he said as he greeted the chairman's secretary. "I need to talk to the chairman right away."

She recalled the recent request by the visiting Harvard professor, Thatcher, who came into the office with a similar urgent request. "Is this about a disciplinary matter?" she said.

"No, not at all," he replied.

"Okay, that's good. I have been afraid lately," she said. "I am frightened to death about what's going on outside. There is so much fear, sorrow, and grief these days. Other than that, everything around here is fine."

"You're not alone," Professor Stein said. "Everyone is equally frightened and uneasy. But I think things will get better, not worse."

"I hope so. Anyway, I now know the sun has thermal nuclear explosions every few seconds, that we have over a hundred billion stars in our Milky Way galaxy, that there are seventy sextillion stars in our universe, and that our universe is only a speck in the allniverse."

"I am impressed! You must have been talking to Jason and Austin about their theories."

"You're right. Jason and Austin gave several of the secretaries and me a full demonstration of your galactic system model. It was great! However, I am angered about the chairman's letter to the Nobel Prize committee. I hope it doesn't affect Jason's nomination."

"What letter?"

"The chairman said he gave you a copy. He said you and Professor Thatcher had determined that you were required to inform the committee of any disciplinary actions of any potential Nobel Prize nominees."

"That's bullshit!" Professor Stein declared. "I have not seen anything!" He stormed into the chairman's office, leaving the startled secretary sitting at her desk.

As he blasted through the door, the chairman jumped back in his chair and hid the whiskey bottle he was holding under his desk. Professor Stein waved his cane at the chairman.

"Do you believe all of that malarkey from Professor Thatcher? About Jason attacking and tossing him to the floor?" Professor Stein yelled. "You must be out of your blasted mind!"

"Hold on, professor. I think Jason's behavior has gone too far," the chairman replied as he set the whiskey bottle down under his desk, unnoticed. "Assaulting the professor in class is one thing, but did he tell you that he also assaulted Professor Thatcher in the gym by delivering a ferocious kick to his rear? Damn it, I am *not* going to allow that kind of behavior here at UCLA!"

Brainstorming The Mystery

"I don't know what you are talking about. Jason has not assaulted anyone, nor did he intend for Professor Thatcher to tumble to the floor. You are allowing that old bastard to pressure you to do something. Even though the world has bigger problems to solve, this will clearly be grounds for the committee to withdraw his nomination in spite of Jason's monumental technical contributions."

"Professor Stein, I'm sorry that you can't see Jason's faults. But he is a disgrace to this institution."

"You know something? You are a blind, whippy drunk, you baldheaded bastard! You and that piglet Thatcher need to go!"

Professor Stein stormed out and headed to the office of the chancellor, one of his closest friends for years, who had the utmost respect for Stein.

35

Helicopter Rescue Failure

Just look at me, Captain Atkins thought to himself. Here I am, sitting on the top ledge of this thirty-story building in the middle of the city at night.

He was trembling. People considered him a big, strong, invincible helicopter rescue pilot. His training allowed him to throw himself into the most dire circumstances. Yet on this day he sat frozen, grasping the roof's ledge.

I was supposed to rescue those little girls yesterday, he thought to himself. Her desperate eyes haunted him. I failed, he thought. I failed every one of them. I couldn't even save my own daughter and wife from that stupid accident last year. I don't deserve to live. "The courageous Captain Atkins." Ha! I don't deserve that name.

He never could have conceived the pain he was feeling, which was far greater than any physical pain. It was excruciating and tore at his soul.

I don't want to live. They're all gone -my father and my mother too. Every time I open my eyes, every moment, I think about my daughter. Her life slipped away. Life is just not working for me.

Tears streamed down his face. His head was throbbing, and his hand was starting to bleed, as he was unaware he was pounding the ledge at his side.

I cannot go on.

He missed his little girl so much. He did not realize how much her loss had been tormenting him.

Helicopter Rescue Failure

But the death of the child yesterday along with most of his squadron made it clear to him that he did not want to live any longer. His body started twitching as he imagined ending his life thirty stories down on the pavement below. He didn't give a damn what kind of mess he would make.

Death will be quick, he thought. I won't have to continue in this unbearable agony. I don't know where my soul will go. I only know that I do not want to stay on this earth any longer. Will there be pearly gates, fire and brimstone, or nothing? Maybe my soul will reemerge in a more pleasurable time and place. I don't know.

Captain Atkins's legs began to quiver and spasm, ready to propel him off the ledge at any moment. His heart was beating like a drum, and his palms were sweating. He could see the city lights as far as the eye could see, and he thought about everyone out there having their own problems. Most were terrified of the mysterious phenomena. Few were aware of or even cared about his existence. They were petrified about the plight of the earth. The cars and streets looked so tiny from up there.

His soul suddenly told his legs to thrust him off the ledge and into the air. As he felt the cool air whizzing past his face, his reflexes caused him to try to grab onto something, anything. His arms and legs began flailing, and he thought, This is it.

He continued falling, falling, faster and faster. Scenes from his life streaked through his mind: his mother in the kitchen making her famous peach cobbler, his daughter playing soccer and running as fast as her little legs could take her. Falling, falling, and more flashes of memory: his father sitting in front of the TV with his fist clenched yelling at the Celtics players, Norman and his wife laughing while on their honeymoon in Tahiti.

He continued falling and saw hundreds of flashbacks. His eyes watered although the wind tried its best to dry them as he fell. Faster and faster . . . He saw the hard, cold concrete coming up to meet him. He tried to get his balance.

Suddenly, he and the concrete collided thrusting him into another spiritual dimension of time and space.

"Daddy, Daddy, Daddy, look at the picture I drew of Mommy," his daughter said as she sat in the den while his wife cooked dinner in the kitchen.

"Honey, that's beautiful. You drew Mommy all by yourself?"

"Yes, and Daddy, look at the red bow I drew in Mommy's hair. Do you think she'll like this picture?"

"Yes, honey, she'll love it. I am so proud of you. You are such a great artist. I am so happy to be here with you, and I will love you forever."

36

Presidential Plane Debate

President Heather Clemson was perplexed and frustrated by all the conjecture and hypotheses about what was happening around the world.

It was now Monday, exactly two weeks since the initial earthquake, and everything was falling apart. The flustered president boarded Air Force One, where she had summoned a select group of expert scientists and religious leaders and her advisory staff to fly back to Washington with her. She was compelled to do this despite the perilous environmental conditions.

The UCLA team informed her staff that they had some breaking news. But they could also share it at the emergency world summit at the United Nations Building in New York the next day. Dr. Braxton's team was finding the claims of the UCLA research team valid. Consequently, the president arranged to have Jason, Professor Stein, and Professor Thatcher to fly back with the presidential team to the White House where they would link into a videoconference with the UN summit. The president was as yet unaware of the turmoil at UCLA regarding the underhanded tactics of Professor Thatcher and Chairman Chapman to disqualify Jason from the Nobel Prize nomination.

Although Jason agreed to fly back with the presidential entourage, he had not been able to contain his anger at and disappointment in the president and those at NASA who had allowed the space station experiment. His only reason for making the flight was that it would allow him firsthand information of any possible contact with the space station and his parents. Jason was unaware how distraught the president was that she had approved the space station experiment that had taken the lives of his parents.

The staff members and guests came together in the small conference room on the DC-bound jetliner. The president tried to console Jason and expressed her deepest sympathy. Jason accepted her apology, but he had yet to fully accept that his parents were lost. He held onto hope that somehow they had survived.

Further, Jason had not yet discussed the details of his theory about the black-hole fragment. He and Professor Stein decided to listen to the others first, as Professor Thatcher would likely challenge anything they said. It took all of Professor Stein's restraint to keep from revealing the disdain and hatred he now felt toward Professor Thatcher. It was unconscionable that Thatcher had generated the letter to the Nobel Committee reprimanding Jason's actions. Stein felt he must keep the information from Jason because his reaction would likely be detrimental to all of them.

The onslaught of arguments, debates, and speculations continued among the other invitees on the plane. Air Force One was equipped with live video communication linked directly into the CNN network. It also had closed-circuit security channels of satellite imagery. The video networks revealed to them much of the horrific destruction of the Hawaiian Islands. Of the over 1.2 million inhabitants of the Hawaiian Islands, most of them were dead. They watched the replay of the devastation and were dazed at the loss of so many lives. Some of them even slipped into a state of near shock.

The enormous ocean swell continued its path of death and destruction, gobbling up smaller islands on its way toward Japan, Korea, and the Philippines. It was not clear whether it would dissipate much by the time it made landfall within the next few days.

Jason was anxious and uneasy since he knew that none of the presidential team members, including the president herself, were aware of the revelations that had emerged from the UCLA research department. He knew they might have even more difficulty accepting the incredible explanations he was about to give them. Nearly every governmental agency and citizen in the world had become involved in intense conjecture and speculation, all focusing on seeking logical explanations of what had been happening.

The president's protocol secretary requested that everyone assemble in the conference room on the plane so the president could proceed with the meeting.

"Thank you for coming," President Clemson said as everyone sat down. "It is imperative we understand what is happening so we can take the proper action. We have recently made some grave mistakes concerning the ISS, and from here on out, we need to act decisively but wisely.

"I want to get some feedback before the conference tomorrow. We need to understand the public's perception. We need to hear from the religious community about the demoralized public. And finally we need to hear from the scientific community about what they have uncovered. However, I want to make this meeting informal and interactive so we can quickly get to the root cause of these devastating occurrences. Let's start. Mr. Bird, what feedback are you receiving from the various sectors of the public?"

"Scattered," said Henry Bird, the press and public relations secretary. "I have compiled feedback from the working-class people, and I will summarize what many of them are thinking. Madam President, most of the beliefs of the American people fall into one of four categories. One, man has flagrantly attempted to alter nature, and this has started to backfire. Two, the whole thing is a hoax or a sophisticated terrorist plot. Three, it is a response from a Supreme Being, and four, there is some unknown planetary or environmental activity taking place."

"Would you elaborate, please?"

"Yes, Madam President," Mr. Bird said. "The first group of people believes we have systematically destroyed our environment, depleted the ozone layer, created excessive greenhouse gases, exacerbated global warming, or exceeded the planet's threshold of tolerable pollution. They believe these are entirely man-induced conditions, and we have irreversibly altered our planet and our entire ecosystem.

"The second group believes that ten years from now there will be no evidence that any unnatural events ever occurred. Everything will be back to normal, and history will show that Mother Nature is alive and well. Many believe the news media has blown this whole thing out of proportion. Others in this group claim they have information that leads them to believe that some of the mysteries are part of a terrorist plot. They believe the government has hard evidence and has elected to conceal it.

"The third group believes there is a spiritual basis for these phenomena. Many believe God, or a Supreme Being, is wreaking havoc because of mankind's wickedness and disobedience to God. And over 45 percent of those surveyed fall into the fourth group, who think something bad is happening to our planet or environment.

"Madam President, the message is this: the American people don't know what to believe and are looking to you to provide the leadership to get us the answers we need."

"Thank you for those comments and that challenge, Mr. Bird," Heather said, thinly disguising her annoyance. "I'll put that request in my in-basket and will have complete answers by close of business today—if that's okay." She crossed her arms and leaned back in her chair. "Cardinal Malloy, what are your thoughts?"

"Thank you for inviting me, Madam President," the cardinal said. "I have sensed some of the same sentiments as stated by Mr. Bird based on the feedback I have been receiving. Many Americans don't know what to think. Some are forming their own explanations and rationale to explain the unexplainable."

The cardinal picked up on the president's annoyance and continued.

"Madam President, you know we are conducting a special vigil at my cathedral. Many prominent religious leaders from all faiths are participating. We are concerned that no matter what occurs, the scientific community will fall back on formulating any scientific explanation for these bizarre events. We believe many scientists will continue with this need for tangible evidence and ignore all others. This is perhaps the nature of man. God gave us free will. Some believe in God while others don't and never will. Some will concoct obscure explanations rather than accept miracles or accept that these mysteries may be God's doing."

"Cardinal, it sounds like you don't have much faith that all of this is explainable or has a credible scientific basis," the president said.

"Madam President, God gives us a choice. We urge you to look deep into your heart and muster all the wisdom and power you have and immediately implement a plan to identify and stop mankind's blatant destruction of our earth. In the beginning, God created the heavens and the earth. Your inaction is allowing man to continue to destroy it. You must issue an order to cease all activities detrimental to our environment and the earth that God created.

"We believe the sky scrolls contain a message for mankind just like the stone slabs that contained the Ten Commandments. We must put our minds together if we are going to decipher the scrolls. It may be something or Someone is attempting to convey a message to us. The Ten Commandments survive to this day after we received them two thousand years ago in a similar manner.

"The religious community has taken the liberty to endeavor to decipher the scrolls. As we speak, we are doing just that. We will keep you informed as to any progress. For now, we are simply praying for our religious, scientific, and governmental leaders to have the strength and wisdom to find the right answers, no matter what they are."

The shaken president listened, thinking it best to allow others to speak their minds. Finally, she spoke. "Who from the UCLA team wants to give us information on this breakthrough? Should I consider it before I issue an order such as the one the cardinal has requested?"

Professor Thatcher started to open his mouth, but Professor Stein jumped in. "Jason Scott will. He has personally done most of the research work at UCLA, which is leading the field."

"Okay, Jason, what do you have to tell us?" the president said.

"Cardinal Malloy and Madam President," Jason stated as all eyes turned toward him. "I believe in God and have been raised in the Christian church. I believe that God has given me the gift to assimilate and evaluate complex concepts, information, and processes, and I believe that God is within me. What I am going to tell you is what God has revealed to me based on scientific knowledge. Furthermore, as stated earlier, we need to keep an open mind to things not familiar to us."

"So what do have to tell us?" Heather repeated.

"I will get right to the point. Based on our research, evidence shows that our solar system was the victim of a rogue black-hole fragment that jarred our planet out of its solar orbit. This fragment had been streaking along a super-galactic orbit traversing our entire galaxy." Jason observed the room filled with startled and suspicious expressions.

"I know this is hard to believe," he continued. "We think it passed through our solar system and streaked right by our planet. The fly-by of this enormous mass imparted an initial jolt to our planet, which we perceived as an earthquake. The mass of the black-hole fragment was likely of an astronomical magnitude, dwarfing the mass of our entire planet by several orders of magnitude. It appears that the sun and all the planets, moons, and asteroid belts were also jolted but remained intact. This black-hole fragment had been moving toward us at close to the speed of light for some time now. We believe it scorched our atmosphere, leaving the sky scrolls in its wake. This is similar to the way meteors, or shooting stars, leave bright streaks in the night sky. The fragment, however, had been traveling thousands of times faster than any known meteor."

Professor Thatcher shook his head.

"We could not see it with our naked eye or with telescopes," Jason continued. "However, several earth telescopes detected something in the sky at the same time the earthquakes and mega-tsunami occurred. As you know, it created a massive sonic boom heard over most of North America and distorted our atmosphere and caused turbulence that resulted in the sporadic lightning and tornadoes. Data is still coming in from all over the world that supports this explanation. We speculate that other black hole near misses may have also occurred in the distant millennia, but never this close to our solar system, and none having such an enormous mass. It is possible that this same infrequent phenomenon has been going on for eons."

The cabin compartment went silent.

Finally, President Clemson broke the silence. In her bewildered state, she reverted by instinct to her legal days. "Professor Stein, how close did this so-called black hole come to us?"

"We are not exactly sure, but apparently close enough to singe our outer atmosphere. I suspect it was within a few hundred miles."

"Could it have hit us?"

"Probably," he said with hesitation, trying to figure out where she was going with this line of questioning. "Actually, yes. If it had collided with our planet, we would have been instantly destroyed."

"You say 'instantly destroyed.' Hasn't our planet been hit by huge meteors in the past?" she challenged.

"Yes, but nothing moving this fast and with such concentrated mass. Look, Madame President, let me be clear about what happened. We believe this fragment was at least the size of the earth, and perhaps much larger. Its mass was orders of magnitude greater—maybe one hundred times denser than our entire planet. If it had hit us dead-on, we suspect everything would have been instantly annihilated."

"I'm having a little difficulty with your notion of the entire earth being instantly destroyed. As far as I can recall, throughout history we have always had some warnings of disaster or impending calamity. We have civil alerts. We have emergency warning systems for even the most dire situations. Sir, people always have an opportunity to duck for cover."

"Madam President, this was a scenario no scientist has ever considered. Let me tell you what might have happened. If a black-hole fragment hit the earth, it would vaporize people within the blink of an eye. People eating, working, and sleeping—kids playing in their schoolyard—would be instantly disintegrated. Every insect, flower, house, ocean, mountain, and continent would be instantly pulverized. There would be no time for warnings or alerts, and no time to pray. People would have no chance to duck, scream, or even be afraid. Every little speck of matter on our planet would be instantly hurled through our solar system and dispersed throughout the galaxy. If that fragment had hit the earth, that would have been the end of our earth and our existence in a flash."

"Okay, okay I get the picture. However, shouldn't we have been able to see it coming? I would expect that our advanced ground telescopes or earth satellite telescopes should have detected something was amiss."

"That's a good point," Jason interjected. "The black-hole fragment was traveling near the speed of light, and we had no means of detecting it in advance. One cannot see light coming toward you, only when it is upon you. We don't know if other fragments are headed our way. This could be the first of a barrage of many spaced apart by days, years, centuries, or millennia. We just don't know."

Professor Thatcher cleared his throat. "With all due respect to my distinguished colleagues from UCLA," he interrupted in a resounding voice, "I believe their premise is completely flawed, Madam President. To be brutally candid I think they are blowing smoke up your rear."

The president's eyes and mouth flew open. She slapped the desk with her hand, and her face turned bright red. "What the hell did you just say? Here we go again. I have had enough of these insults."

Cardinal Malloy, Mr. Bird, and the rest of the president's staff looked at Professor Thatcher in shock as he peered back at them through his thick glasses.

"What the hell do you want to say before I ask my security officers to take you to the back room, Professor Thatcher?" the president growled.

Jason and Professor Stein had been expecting Thatcher's opposition.

"We at Harvard have some of the foremost experts on these matters," Thatcher continued. "First, UCLA speaks of a black hole as if it were a single entity or body. Any credible group familiar with astrophysics knows that a black hole is a system that consists of a concentrated mass of extreme density, likely formed by an orderly collapse of a star over a long period and surrounded by interstellar bodies. A black hole is not simply a rock flying through space, as the UCLA brats would have you believe. I have been doing research on black holes at Harvard from their first discovery just over twenty years ago."

"Excuse me, sir," Professor Stein said. "None of us 'brats' at UCLA said the object was a black hole whizzing by. We indicated it appeared to be a *fragment* of a black hole. We know that your research team is falling far behind ours by injecting global environmental factors, but that is no reason to discredit our team."

Professor Thatcher began to rebut, but President Clemson quickly stopped him.

"Gentlemen, gentlemen, hold on for a moment. We're not here to debate the semantics of black holes. What we believe is that something whizzed by the earth and nearly knocked the hell out of us. I don't care if it was a big black hole, a black-hole fragment, a white hole, or an asshole. Something happened! You men need to get your acts together and work on explaining what happened. What we need is to focus on what are we going to do from this point forward. I hope that is clear because I don't want to hear any more of these childish arguments, especially during the emergency world summit tomorrow. Now, if you'll please excuse me, I'm adjourning this meeting. I need some time to digest what I just heard from you folks," she said in a weary voice.

Professor Stein gripped Jason's arm and practically dragged him to the other compartment before Jason reacted. Jason's heart pounded like a jackhammer and a vein twitched in his neck. They sat down in adjacent seats.

"Jason, listen carefully," Professor Stein whispered, leaning closer to him. "Calm down. I didn't want to bring this up, but I talked to the university chancellor yesterday. He said he received a call from the dean at Harvard concerning a big problem. They will not be accepting Professor Thatcher back there because of his terrible rapport with the students and his damaging history of incomplete projects. The chancellor informed me that UCLA is having the same problem here. So the professor will be out of a job real soon. The chancellor also said that Professor Thatcher does not realize that Harvard sent him to UCLA because of his poor teaching ratings. As it turns out, students were able to achieve at Harvard in spite of him rather than because of him. He thought he was coming to UCLA to upgrade our academic level in the field of cosmic sciences. So don't worry about his angry outbursts. He's a goner."

Little did Jason and Professor Stein know that Professor Thatcher was imposing himself on the president and Cardinal Malloy. He was trying to convince them to abandon the seriously flawed UCLA research in favor of a takeover by his prized Harvard team. He suggested to the cardinal that he would join him in rejecting any rash, unproven scientific ideas and suggested they advocate a more cautious perspective, claiming that Harvard and his UCLA team had already initiated research that advocated descoping scientific intrusion into the earth's natural resources. Their research proved that the earth's ecosystems were at a breaking point and that man-induced environmental changes were becoming unstable. Thatcher knew this slanted statement aligned with the cardinal's viewpoint.

"On top of this, I have some regrettable news," Professor Thatcher said. "I feel compelled to inform both of you that charges are being filed against Jason Scott for aggravated assault. He attacked me in the classroom and in our locker room. He is a bright student, but he is not stable, and I fear for my well-being around him."

The president and cardinal were stunned by these accusations.

As Air Force One approached Washington DC, the attention of those onboard was drawn to CNN.

"A young student attempted to shoot a NASA official as he was entering his limousine after participating in the summit in Vandenberg Air Force Base," the reporter said. "The young woman, a student at Cal State University at Long Beach, has been identified as Britney Scott, the daughter of the husband-and-wife astronaut team stranded aboard the International Space Station that is currently lost in space. She is being held without bail in the Los Angeles county jail."

Jason was staggered.

"Don't worry, Jason," Professor Stein sighed. "We'll head back to LA and take care of your sister."

37

The Traumatized Girls

After making some phone calls, the White House excused Jason so he could fly back to California, hoping they could remedy his family situation quickly. If so, this would provide him additional time to focus on his research at UCLA. He would have to rush to prepare for the critical second summit slated for the next day. Jason hoped to gain alignment with the other scientists, including the Harvard research team.

Back in Los Angeles, Jane had been spending considerable time in her apartment. She stood on her balcony gazing at the sky scrolls, oblivious to the breaking news about Britney. Even though the world was in near chaos, Jane was aware only of her own terrible dilemma.

Manuela had noticed that Jane had been acting strange lately and was avoiding contact with Jason. When she finally confronted her, Jane broke down and told her that she had missed her last period and did a home pregnancy test.

"What are you going to do?" Manuela asked. "Have you told Jason?"

"No!" Jane said. "It's not Jason's baby."

"Wh-what did you say?" Manuela said. She took a step back, staring at Jane. "Girl, what are you talking about?"

"Jason and I practiced safe sex. This is Luther's baby."

"You slept with Luther??"

"No! He raped me when we were in Mexico."

"Oh, my God!" Manuela clamped her hands over her mouth and stumbled back against the wall. "Why didn't you say something? You need to take care of this! You have to get an abortion and tell Jason what happened!"

"I can't get an abortion. That goes against my whole being. I have a living life inside of me. It's part of me now. If I told Jason, it would devastate him. He has suffered so much already with the loss of his parents."

"Well, what the hell are you going to do?"

"I don't know. Die, I guess!" Jane began to sob.

"Hey, Jane, take it easy." Manuela rushed over and put her arms around her. "Are you sure you're pregnant?"

"I am certain."

"Why don't you go to the clinic to be absolutely sure."

"I am absolutely sure!" After a moment, Jane added, "Maybe I'll get checked out at the clinic anyway."

"I'll go with you."

"No, no, I'll go by myself. I need to be alone."

Jane braved her way to the clinic later that day. Her suspicions were confirmed, making her even more distraught. What was she going to do?

38

Retrieved Sister and Letter

Soon after arriving back in Los Angeles, Jason, along with the government personnel got his sister Britney released from jail. They also managed to get the assassination charges suspended based on grounds of temporary insanity and got her situated in a low-key psychiatric ward of a private hospital for her to recover.

When things were under control, Jason joined his research team and continued collecting data and rerunning his galactic model. He planned to collaborate with his colleagues and other researchers from around the nation to collect their views, observations, and findings. He had been so engrossed in his work and the traumatic events in his life that he had not seen or talked to Jane recently. As soon as it occurred to him that she had not returned any of his calls, he dropped everything and headed to her apartment to see what was going on.

When he got there, Jane was gone. He confronted Manuela, asking what was up with Jane and why she hadn't returned his calls. After Jason's intense probing, Manuela finally broke the news to him about her pregnancy.

"Why didn't she tell me?" Jason yelled. "Is she okay? Why didn't she call me? Where is she now?"

"Dammit, I don't know, Jason," Manuela confessed. "She might have gone to the campus clinic." She didn't have the nerve to tell him that he was not the father.

Alarmed, Jason ran out of the apartment and over to the clinic, but Jane was nowhere to be found. Still, he continued his search. Eventually, he found her at her mother's house in Redondo Beach.

Jason asked her if there were something she wanted to tell him because he had just talked to Manuela. Jane, assuming Manuela had told him everything, sighed, hung her head, and said, "I'm pregnant with Luther's baby."

Shocked speechless, Jason braced himself against the door. "*Luther's* baby," he said in a low, unsteady voice. "*Luther's* baby . . .?"

"Yes," Jane confirmed, still avoiding his gaze. "When we were in Oaxaca, he raped me."

"Luther raped you!"

"Yes, after I took too much Valium, he took me to our guest dorm room and raped me."

Jason couldn't believe what he was hearing. Memories began racing through his mind. He recalled when he first saw Jane at Ackerman Hall with Austin, the first time they made love before her trip, his romantic marriage proposal at the airport, and her being at his side when they flew to Vandenberg. Luther had committed the ultimate violation of this woman Jason loved more than anyone in the world.

Jason clenched his fists and his lip began to quiver. "I am going to kill that bastard!"

"That won't solve anything, Jason," Jane said as she grabbed his arm. "I am carrying a baby inside of me, and Luther is the father. Killing Luther won't change that."

"Jesus Christ!" Jason pulled away from Jane and stomped around the room, frustrated and confused. "We have to do something about the baby."

"What are you saying?" Jane said as tears trickled down her face. "We can't destroy a life. We both believe that abortion is murder, no matter what the situation. I am carrying a God-given life that has every right to exist. I love you with all my heart, and that will never change. But I am carrying Luther's child, and we have no right to decide the baby's fate." Jane tried to convince Jason to not worry about Luther, as they had the pregnancy to deal with.

"Jason, please listen," Jane pleaded. "I'm not sure how you will feel about the baby. I know that I never want to see Luther again in my life, but the baby only has me to depend on. I know you might never accept this baby, that you will likely resent him or her and perhaps even me."

"Wait a damned minute," Jason said. "This is all happening way too fast. All I know is that I love you very much and I want to marry you and have a family together. We have to figure out a way to make that happen. I can't believe this is happening." Jason paced around the room with his hands on his hips and then suddenly punched the door. "I have to get away from here. I have to leave now so I can figure out what to do." He stormed out, leaving Jane alone on the balcony.

Jason searched for Luther despite Jane's pleas not to do so. Ignoring the sporadic lightning overhead and the unpredictable tornadoes, he embarked on his manhunt. After searching everywhere, he walked through the campus and passed the clinic. Exhausted, he decided to go in and ask for advice.

He made his way to the main desk and asked for Dr. Michael Larson. The portly doctor and Jason had a past history when Dr. Larson was an intern practicing internal medicine at the clinic. He had accepted this position as the first step toward eventually opening his own practice.

"Dr. Larson is busy seeing patients now," the receptionist said.

"I'm a good friend of his, and there's an urgent situation. It will only take a minute."

"I'll see if I can schedule a few minutes of his time with you."

Moments later, she ushered Jason into Dr. Larson's office.

"What an honor to see you, Jason," Dr. Larson said cheerfully as he stood up from his desk. They warmly greeted each other with their customary handshake and pat on the back.

"Thank you for the kind words."

"I can only imagine what you've been going through with this whole earth-mystery ordeal and now your parents."

"Yes, this has been the most difficult time in my life."

"I'm really sad to hear that, Jason. But I know you're strong and will pull through. These events have traumatized all of us. Do you realize most of my staff has left because of all the occurrences?"

"I did notice that it looks vacant around here and throughout the whole school."

"And I'm sorry about your parents. Is everything else okay? Is there anything I can help you with?"

"Yes, there is." Jason hesitated and took a deep breath to conceal his frustration, anger, and confusion. He was reluctant to tell Dr. Larson about Jane's pregnancy. The news was still swirling around in his head. "Dr. Larson, Jane was here earlier today and found out she is pregnant."

"I was not aware of that."

"You know we are engaged now."

"Well, let me say congratulations."

"Thank you."

"So what do you need?"

"I would like for you to run some tests on the baby and me."

"DNA tests? I presume by your request that you two are concerned about the baby's health?"

"Ummm, yes," Jason said.

"Sorry, but to do that test we must have the mother present and she must give her consent."

"Dr. Larson, I only want to compare DNA between the baby and me."

"That is a usual request, since we are only allowed to perform limited testing."

"I would really appreciate whatever you can do. That would help relieve me of any more worry."

"I assume you are concerned about genetic defects?"

"Umm, yes." Jason nodded. "That is what I'm concerned about."

"Jason, for any type of prenatal or paternity test you need the mother's permission."

"Doc, this is really urgent and you know with all that is happening to the environment things are not normal. Is there anything you can do?"

"Jason, to do what you are asking you need a blood sample of the mother at the end of her first trimester -ten weeks of pregnancy. She has only been pregnant for a short time."

"Isn't your clinic doing some state of the art research?"

"Yes, but it is research only. However, with our advanced DNA SNP microarray technology we might be able to examine her and the baby's pregnancy condition. And, we do have a blood sample from Jane. So let me think about it."

"Doc, it would mean so much to me if you would at least give it a try now."

"Okay, I'll have the nurse pull Jane's chart and draw some blood from you. We will use our research work and run the parent/child DNA compatibility profile for the father only. We really need Jane's permission but given the circumstances I'll try. With our new technology, perhaps we can have preliminary results in about an hour. That is, if no more people have abandoned us in the lab. I'll have the nurse show you to the examination room, and I'll see you shortly." Dr. Larson shook Jason's hand before summoning the nurse.

The nurse led Jason to the lab and drew a blood sample. She then asked Jason to wait in the lobby. He knew time was running out to resume his research simulations, but this was more important. He paced around the room, gripping his fist and thinking about how his life was crumbling.

The nurse returned just over an hour and summoned Jason to the examination room again. Dr. Larson looked at the results for a few minutes in silence.

"Great. Everything looks great, Jason."

"What do you mean, 'great'?"

"We ran the DNA tests to check for inherited disorders. The baby looks fine, and you and Jane have no inherited disorders."

"What do you mean, 'the baby looks fine'? Can you tell if the baby is mine?"

"Of course! Even though this in not a paternity test we can tell. " Dr. Larson said. "The baby is yours. The DNA shows a perfect match. You are definitely the father."

"The baby is mine?" Jason asked with joy.

"Of course! There was a perfect DNA correlation," Dr. Larson said, puzzled.

"Jesus Christ," Jason shouted. "Are you sure it's mine?"

"Yes, yes, yes," Dr. Larson said. "I am absolutely sure the baby that Jane is carrying is your child."

Jason was unable to contain himself. He gave Dr. Larson a big bear hug, lifted him off the floor and spun him around. His loud, unrestrained laughter astonished the doctor and the nurse. Jason executed two pirouettes before dashing out of the clinic to tell Jane the good news before she did something foolish.

When Jason arrived back at Jane's parent's house, he rang the doorbell and called for her, but no one answered. He peered through several windows and even climbed to the second floor to peer through her bedroom window. Nobody was home. He then decided to go back to Jane's apartment.

When he arrived there, he double-parked, ran to her building, and rang the doorbell several times. When no one answered, he used his key and charged up to her apartment.

Jason walked in and called out for her. As he approached her bedroom, he found the door slightly ajar. He called for her again, but there was dead silence. His heart began pounding as he sensed something was very wrong. He pushed open the door and prepared himself for what he might find. Trembling, he glanced around her room.

Something caught his eye. He walked over to her bed and found a note addressed to him. With shaking hands, he opened the envelope and began to read:

Jason, when you get this message, I will be gone. I love you with all my heart and soul and I always will. I cannot bear to make you suffer with having to raise a child not fathered by you. I would not be able to live with myself if I did anything to destroy or cast my child off to another person or situation. I know you have already lost so much, and it is not fair to bring the burden of this baby into your life also. I am leaving and you will not ever be able to find me, but I want you to know that I will always love you with all of my heart. I am going to name my child Jason Jr. anyway.

Good-bye

Love,
Jane and Jason Jr.

Jason collapsed to his knees in the doorway.

"Please, God," he cried out. "Please don't let them go."

He could not fathom how the beloved person he saw every day, whose existence was inextricably linked with his, could be gone forever. The brightness of her eyes and the sound of her tender voice, so familiar and dear to him, would be no more.

Such reflections would haunt him days to come, and as time relentlessly marched on the reality of the situation would not subside, allowing the anguish, grief, and misery to endure.

39

Summit in New York

The next day, all the distinguished world leaders were chauffeured to the United Nations Building on the east side of Manhattan for the second emergency summit. After disembarking from the black limousines, the top scientists, religious leaders, and UN ambassadors made their way through the swarm of spectators who had gathered outside and filed into the building.

Scientists from all over the world had been working nonstop to corroborate or refute the UCLA findings, and they were planning to reveal their findings at the summit. As the participants entered, they were ushered to special tables to accommodate the large number of attendees. History was in the making as the fate of humanity hung in the balance.

Dr. Braxton had received a draft plan from the White House to review. It detailed all of man's scientific and environmental activities deemed destructive to the earth and its ecosystems. Around the same time, he received another report that corroborated the UCLA research team's astonishing hypothesis. He was of the mind-set that this was his time to shine and show the world what he and his NASA organization could accomplish in a severe crisis

Dr. Braxton had arranged for a high-tech interactive computer system to be placed in the center of the circular chambers with surrounding seats for each of the two-hundred-plus member nations. (The chamber has a total seating capacity of over eighteen hundred.) He had also positioned large video screens throughout the chambers and a computer system at the center table like that of a sports arena.

UN Secretary-General Bay Loy called the summit to order and requested the spokesperson for the scientists to present their confirmed findings. She then asked Dr. Braxton, who had been working closely with Professor Stein and the UCLA research students, to come forward.

"I have some extremely disturbing news to report," Dr. Braxton began and turned to look at the secretary-general. "Using all available resources, including ground and space telescopes and ground radar, we have confirmed that the research of Jason Scott from UCLA is correct. Our planet did indeed experience a near miss from an enormously dense cosmic or celestial body. These findings have been verified by NASA technologists and institutions around the world and indicate that the earth was jarred into a different elliptical orbit around the sun."

"Will this affect us?" an impatient representative from Italy blurted out.

"Yes, to be candid, and I'll get right to the point," Dr. Braxton asserted. "It will be catastrophic. I am going to have my NASA director of astronomy, Dr. Bill Zelman, come forward to explain exactly what the impact means to us and what we need to do."

A current of disbelief rippled among the members. The room began to hum as the representatives murmured to one another in an effort to confirm what they had just heard.

Dr. Zelman was one of NASA's most highly regarded experts and was known as the Einstein of space exploration. He had actual experience working on numerous space programs, including the recent Galilee II and Neptune probes exploring the outer fringes of the earth's solar system. As he walked forward, he cuffed his right hand that held his presentation material. The audience chamber members were by now very anxious.

"Let me begin with what we know," Dr. Zelman said. "I don't need this material," he held up the packet in his right hand, "or visuals for what I'm about to tell you. I am going to speak from my heart for a few moments.

"Our situation is ominous. We have been working closely with Professor Stein and Jason Scott of the UCLA research team and have confirmed their findings. As you may or may not know, their work is monumental.

"Let me tell you what is happening. The earth normally moves in a nearly circular orbit around the sun once every twelve months. Its axis is tilted about twenty-three degrees. It is this tilting that accounts for our seasons."

He approached the computer technician operating the video display terminal.

"Let me interject some fundamental notions to prepare you for the conclusions to follow," he said. "Please look at the computer monitor Dr. Braxton has set up. Every 365 and a half days, the earth completes its orbit of the sun. During that time, the weather over much of the world changes in a regular pattern, known as seasons. Year after year, spring, summer, fall, and winter follow one another. Spring always begins around March 21 in the northern hemisphere, and fall starts on the same date in the southern hemisphere."

The computer monitor depicted the earth moving around the sun and showed the change of seasons.

"The earth has seasons because its rotational axis is tilted. You can see on the screen the North and South Poles are tipped at an angle of about twenty-three and a half degrees. The earth always leans in the same direction as it revolves around the sun. This is normally the key to our seasonal and temperature changes. The sunlight hitting the northern and southern hemispheres changes seasonally as the planet orbits the sun. On about June 22, when the northern hemisphere tilts directly toward the sun, countries, such as the United States, enjoy the first day of summer. This is the longest day of the year, when the sun's rays are most direct. Meanwhile, it is the shortest day in the southern hemisphere, as the South Pole points directly *away* from the sun. If the earth had no tilt, we would have no seasons."

"So what?" whispered the restless member from Chile to his colleague seated next to him. "So we now know why we have summer and winter. What's the point?"

"What I am going to tell you now is going to sound technical, but it is imperative that you understand. Please pay close attention," Dr. Zelman continued. "During the earth's normal yearly journey around the sun, we experience both a closest point to the sun and a farthest point. Because of the recent worldwide events, we are now following a far more elliptical orbit than usual. Generally, in our solar system, the farther away from the sun a particular planet is located, the cooler the planet. Some closer-orbiting planets like Mercury and Venus are extremely hot and eruptive while the most distant planets or solar bodies like Pluto and Neptune are extremely cold and quiescent. Life as we know it cannot exist in either extreme.

"Recently, the geoscientist John Hopping developed the notion of the 'survivable zone.' This zone represents the farthest and closest distances from our sun—or any sun—in which life can exist. For our sun, the zone is between seventy-nine and 140 million miles. Presently, astronomer Darnel Wilson and his colleagues at Northeast State University have been studying elliptical orbits using their Synthesis 4 computer model. They have input our new orbit data and confirmed that our earth's orbit has both elongated and shifted until it is now about 14 percent closer and 18 percent farther at its extreme positions. Previously, the average surface temperature of our planet was fifty-eight degrees Fahrenheit. However, now at the closest point, the temperature will average just over 150 degrees Fahrenheit. Consequently, most if not all of the ice in the Arctic and Antarctic regions will melt, and most of the continents will become flooded. The atmosphere will be mostly steam with little oxygen. What this means in simple terms is that all human life and almost all other life forms on Earth will be destroyed."

Everyone in the room gasped, and soon pandemonium broke out. After several failed attempts to restore order, Dr. Braxton directed one of the security guards to fire his rifle toward the ceiling, which created a deafening echo throughout the chamber. Everyone ducked to the floor, and Dr. Braxton grabbed the microphone.

"Please get back to your seats! You need to hear our options."

The UN secretary-general and the other leaders stumbled back to their seats.

"We were just as shocked as you are," Dr. Braxton said. "However, we have been up for days on end consulting with experts from all over the world trying to come up with solutions, not just give up and pray that things will get better. We have brainstormed many possible options and believe we are left with at least one viable option. Let's allow Dr. Zelman to explain things before we go off half-cocked."

"All is not lost," Dr. Zelman asserted as he stepped forward again. "The last option I will discuss is the most viable. In summary, the first option we considered was accepting that billions of people will likely die but that many could be saved for a while by constructing underground structures. This would require developing protective houses and even whole neighborhoods for selected people. Another option is to accelerate construction of the underwater ocean village structures and send selected people to inhabit them. Another would be—"

"Enough of these absurd options!" the leader from Canada shouted. "Get to the point! The whole world is about to be destroyed. Tell us about your viable option."

"Calm down," Dr. Zelman pleaded. "Dr. Walker, are you ready?" He glanced over at the table where his scientific team sat. "Sonia, are you ready to discuss the idea your team has been focused on?"

"We're not completely ready, but I guess it wouldn't hurt to reveal what we have so far."

Dr. Sonia Walker was one of the brightest new leaders to emerge from the NASA ranks in a long time. Every member of the UN had complete respect for this young African-American woman with demonstrated technical strength, strong leadership ability, and deep wisdom. Along with her short stature and almost boyish haircut, she had a booming voice.

One of Dr. Walker's strengths was her tenacious pursuit of a goal once it was established. She was driven by a need to overcome the stigma and disgrace brought to her family by her infamous late grandfather, who was misunderstood as a militant focused on overthrowing the government. She strived to rectify this misconception by devoting her career as a loyal servant of the government and of the people of the United States. She was the right person for this demanding endeavor, and she would endure until the end.

"My name is Dr. Sonia Walker," she said after lowering the microphone. "I am the team leader for what Dr. Zelman describes as the only viable option. I will get right to the point."

Everyone was on the edge of their seat.

"We and the UCLA team think we have a solution to the problem. It is incredible, but possible. We think we can realign the earth into its correct orbit. We can do this by applying a large impulse force." She gazed around the room at the sea of mystified faces. She took a deep breath and said, "We believe we can generate a force large enough to move the earth."

The room started to buzz again.

"Do you mean you are going to try to move the *entire earth*?" the representative from Switzerland blurted out.

"Yes," Sonia replied, squeezing the microphone with both hands and gritting her teeth. Her firm voice echoed throughout the chamber.

With arms crossed, the baffled Secretary-General Loy asked, "What would it take to make that happen, may I ask?"

"Megatons," Sonia said. "We will have to explode several hundred nuclear bombs, each having a magnitude of two to three hundred kilotons of TNT, at a preselected region on the planet."

All jaws dropped. The entire room started to churn, and the members looked at each other in disbelief.

"I don't think there are that many nuclear bombs in the entire world," the leader from Russia said, striving to restore logic to the proceedings. "And worse yet, you probably would kill us all as well as destroy the entire planet in the process."

"Please let me finish before you jump to conclusions. We estimate that several hundred nuclear bombs detonated within holes, or silos, strategically located on the continent of Australia would provide enough of an impulsive force to realign the earth into its proper orbit. The detonations will have to take place within a brief time frame. They will have to occur 211 days from now, when the earth is in the right position to minimize the force needed for realignment. The exact impulsive force required is undergoing verification. My team has investigated the viability of this endeavor and determined that it is feasible based on several simulations, including those conducted by Jason Scott's celestial model at UCLA. In theory, several government and research institutions have shown it will work. We are continuing to investigate all the downside ramifications but believe we have solutions for each concern that might arise. However, I must reemphasize that the theory is solid, and a plan to mitigate all the adverse consequences must continue to be refined. Nonetheless, the actual implementation is beyond my control and will be discussed in a few moments."

"There must be another location rather than my country, Australia!" Dr. Braxton said as he stood up in a rage, pointing a threatening finger at her.

A barrage of questions bombarded the young woman, and soon, order was lost. The secretary-general ordered all microphones be turned off. "Please, one question at a time," she said. "Raise your hand, and I will reactivate your microphone."

"What theory are you using to arrive at such a bizarre solution?" the delegate from India asked.

"If you will allow me," Sonia said, "I will give you a short technical description of the plan. I will write out the relevant mathematical variables on this pad in front of me and have them projected for all of you to see."

"Why not?" the secretary-general said. "I feel helpless at this point."

"Thank you," Dr. Walker said. "The necessary force, for the most part, is dependent on the mass of the earth and the mass of the sun. It also is dependent on the change in the distance of the earth from the sun before the misalignment and the magnitude and duration of the detonation force. The magnitude of the detonation impulse force is given by:

$F*t$ = m*[square root(MG/r1)+ square root(MG/r2)] where:

F= magnitude of the impulse force

t= duration time of the force

MG= Mass of the sun times the universal gravitational constant

R1= Distance of the earth from the sun before the jolt event

R2= Distance of the earth from the sun after the jolt event."

"Hold on," the secretary-general interjected again. "This sounds like too much mumbo jumbo. Are you serious about this idea? For one thing, what would happen to all the people in Australia? Wouldn't the nuclear fallout kill us all? Wouldn't this blow up the earth?"

"Let's take it one step at a time, please, Madam Secretary. We will have to evacuate all thirty million inhabitants from the Australian continent and situate them in other countries."

"Let's stop for a moment," the secretary-general said. "I'm not sure I'm getting all of this. We appreciate your quick response to this world crisis and the superb job you have demonstrated thus far. But we need to catch our breaths and digest what your team is telling us. So let's do this. Let me caucus with the Security Council leaders for a few moments. We'll take a fifteen-minute recess."

Secretary-General Loy huddled with the Security Council and Dr. Braxton. After several minutes, she made an announcement over the speaker system, and the people scurried back to their seats.

"Given the magnitude of what has occurred, we need to regroup," the secretary-general said. "To deal with the flood of information and stunning revelations that have been disclosed here requires that we grasp the information. However, we cannot think of any other agency or organization equipped to handle it any better than your team. Therefore, we will do our best to put some structure into these issues unless someone else has a better idea. With that said, I am going to assume this situation is the top priority for the entire world, and we will request that every necessary resource and supporting entity be made available to us. We request that each of you representatives make this happen. We are invoking the emergency powers of the UN.

"Clearly, what faces us is a technical situation beyond the scope of anything we can imagine. The social and logistical difficulties associated with a problem of this magnitude may dwarf the task's technical difficulties. What we propose is to reconvene here in two days, with the same participants. This will give everyone a chance to digest what has occurred. We would like this to be only a nightmare, but we are all wide-awake, and this is real. We have asked that Dr. Braxton continue to head all activities. We request that you give him and his team your full support and cooperation.

"We will reconvene here early Thursday morning with a closed-door session between the team and designated leaders from the UN Security Council. That should give the Braxton team enough time to verify its findings and develop a plan of action. Does anyone have anything to add?"

There was silence throughout the chambers as everybody exchanged glances.

"Wait a moment," Cardinal Malloy said as he stood up. "We are moving too fast. We need to pause and think for a moment. Have you ever heard of dinosaurs?"

Everyone in the chamber stared at him. What now?

"Our scientists tell us they became extinct over seventy-five million years ago. They now claim that man came into existence four to five million years ago. In fact, the great dinosaurs and man never coexisted. Scientists purport that a great meteor struck the earth seventy-five million years ago and wiped out all of the great lizards along with nearly all other living land animals. It took another sixty million years for man to emerge, according to scientists. Therefore, is it likely that if the meteor had never hit the earth, man may never have evolved? Do we owe our lives to this great meteor?

"Similarly, if Noah had not rescued his family, the animals, and himself during the great flood, life might be completely different, perhaps even nonexistent. In other words, if it were not for the meteor and the great flood, man might not exist today, if at all. Thus, it might be a Higher Power that inflicts havoc like meteors, floods, or black-hole fragments on this planet to make way for new life, such as us. With that said, another option is to let God's will be done and allow life to continue as God and Mother Nature determine. If this is the case, we should consider not interfering at all."

The cardinal's words were met with no applause or commentary, only silence. Finally, Secretary-General Loy spoke.

"Well, I want to thank you, Cardinal Malloy, for those thought-provoking comments. I think you have a compelling point. We need to pause for a moment and consider our plans from every angle." She looked around the room. "We will reconvene this body on Thursday as planned. We will consider all of these options, including the one just mentioned by Cardinal Malloy to do nothing. Thank you for your efforts. The meeting is adjourned."

As the representatives exited the room, the secretary-general sat thinking to herself, I doubt many religious people will go along with this "do nothing" approach. However, there is absolutely no way the scientists can firm up their findings and digest all the

Aphelion

issues in two days and return here with anything
credible.

40

Earth Realignment Project Announced

Over the next two days, all normal world activities were suspended as people around the world fell into shock and terror over their plight. Many walked around in a state of utter hopelessness, resigned to the fact that the world as they knew it was going to end and there was nothing they could do to stop it. Another segment was frantically hoarding food and supplies as if they were preparing for a war.

Dr. Braxton's team was frantically working on the colossal burden placed on them. They had somehow managed to encourage a great many people in the world to support their every effort.

On Thursday morning at 8:00, Dr. Braxton and his entourage reconvened at the United Nations Building. They were to brief the Security Council with their airtight plans while nearly everyone on Earth tuned in to watch the summit session. Dr. Braxton and his team were fortunate to secure the presence of Jason Scott via teleconference despite his despondence over the loss of his parents, his sister's recent incarceration, and his ongoing search for his fiancée. Dr. Braxton promised he would put the full resources of the national government at his disposal while he continued to support all technical efforts and signed into the summit.

Dr. Braxton began the summit with the caveat that they were still collecting data and that there were numerous parallel activities underway to confirm their findings and recommendations.

"We now know what we must do," Dr. Braxton said with conviction. "We believe we have sufficient corroborative information to recommend action to the governing body. We must convey several critical aspects, so please allow us time to explain our findings before posing any questions. We will get right to the point.

"The world is now faced with the greatest challenge ever to confront mankind—the imminent end to our species and our civilization. We have the option of doing nothing or we can do something about it. It will take unprecedented cooperation from all the nations of the world to avert doom. We must take the necessary steps and strive for unparalleled trust, openness, and cooperation. I will first summarize what we believe happened to us and then proceed from there.

"I would like to direct your attention to the screen. The UCLA research team headed by Professor Joseph Stein provided the simulation and backup information. Jason Scott was key in assembling this information in spite of his recent personal tragedies.

"The decision to arrive at the right option as posed at the end of the last summit was not a difficult one for my team or for me. A summary of viable options is limited to these alternatives: one, detonation for the realignment of the earth, or two, let nature run her course and hope the world survives the extreme climate shifts.

"We have evaluated the findings and have arrived at a unanimous decision: we will detonate. We will immediately discontinue all other conflicting options, including the humanitarian option suggested by Cardinal Malloy to let nature run her course.

Further, we have dropped the option by Professor Duncan Thatcher and the Harvard team to continue studying and evaluating the impact on the environmental climate. This option only serves to detract from the validated UCLA hypothesis that we will now vigorously pursue."

The assembled leadership nodded in agreement. However, they had no idea what the Respect Mother Nature Society (RMN), headed by Lamont, Thatcher's research student, was planning. Thatcher was not going to sit idly by while their option got overlooked.

"Please, direct your attention to the video screen," Dr. Braxton requested. "The UCLA simulation shows the earth as it was on March 14, orbiting around the sun. As you can see, in slow motion, an object of extraordinary mass passes near the earth, close enough to scorch our atmosphere, leaving in its wake what we call the sky scrolls. This near miss jolted the earth, which we mistook as a worldwide earthquake and sonic boom. It churned our atmosphere and created the widespread lightning and tornadoes. This object pulled on the ocean, as does the moon, causing the ocean to swell and create a mega-tsunami that enveloped islands in the Pacific, including the Hawaiian Islands.

"Many other events occurred when the object passed our planet. At that time, nobody linked them together. The near miss pulled the earth into a slightly different orbit around the sun. We believe the dense mass that started all this had to be enormous. Materials from black holes are the only ones known to be associated with extremely concentrated mass. Even though it is believed that only X-rays escape from black holes something happened. Who knows for sure, but some type of incredibly dense mass affected our planet. Its velocity upon passing the earth was close to the speed of light, 186,000 miles per second, which explains why we never detected its approach.

"As you can clearly see from the video simulation, the black-hole fragment affected the moon's alignment with the earth. Scientists are quantifying the extent. Astrophysicists from the Arecibo Observatory in Puerto Rico and researchers from the Keck Observatory in Hawaii—before its destruction—have confirmed our findings. We also have confirmation from the Hubble telescope orbiting our planet. It appears that the streaking object obliterated three of our hundreds of earth-orbiting satellites."

The audience sat frozen in their seats, glued to the large-screen monitor.

Braxton continued. "Because of the imminent catastrophic consequences to the earth in the wake of this object, we are not going to focus on describing the cause of our situation. We must turn our attention to whatever corrective actions we believe are possible. Since the entire world is in peril, it is critical that we act decisively. Therefore, we will curtail this discussion and have Dr. Sonia Walker and her team present the actions we must take. I am encouraged by and confident in their plan. I suggest you hold your judgment until you hear everything she has to say."

The world watched on.

41

Key Computer Lab Attack

Jason was in a building on campus where a high-tech video monitor was available. Although still emotionally fragile, he managed to hold himself together to support the summit via teleconference feed. His team was keenly aware of his difficulties, but they also knew he was determined to use his research to help the world in its hour of need.

Austin and some of his colleagues were busy in the research lab while the reconvened second summit was in session. They needed to verify their data so they could rerun the model simulation to evaluate the altered earth orbit and then transmit their results directly to the summit team.

As they were about to begin the simulation, a radical group of protesters, allegedly from the Respect Mother Nature Society, headed by Lamont (Thatcher's student), stormed into the lab wielding clubs and baseball bats. They quickly overtook Austin and his colleagues.

"YOU ARE DESTROYING OUR PLANET!" Lamont shouted. "We have to stop you and your flawed research. We are answering to a Higher Power, and our intention is to destroy this place. You scientists are going to screw things up again. Nobody in their right mind believes our planet was sideswiped by some invisible rock!"

He rushed over to the computer monitor where one of the students was sitting, raised his metal bat high in the air, and brought it down on top of the monitor. Glass flew in all directions, and the student tumbled to the floor and scrambled away.

"Are you crazy, fool?" Austin yelled. "What the hell are you doing?" He rushed over to the leader, grabbed the bat, and wrestled him to the floor. As Austin was about to overpower him, Lamont crept up behind him, raised his bat, and delivered a devastating blow across Austin's shoulder and neck, knocking him to the floor, causing his gold chain to snap apart. Austin fell to the floor, groaning. Lamont stood over him as Austin was on his hands and knees trying to recover. He raised his bat high into the air again and slammed it across the back of Austin's neck. There was the sound of crushing bone. Austin collapsed to the floor unconscious.

The other horrified research students escaped down the hall. The remaining radicals from the Respect Mother Nature Society went on a rampage, smashing all the equipment in the lab. They continued their frenzied destruction of all the monitors, computer systems, and vital storage records like rabid dogs, as Austin remained motionless on the cold floor.

42

Summit Reconvenes

Back at the summit, Dr. Sonia Walker returned to the podium. The room was filled with a sense of hopelessness. She and all the leaders were oblivious to the tragic events that were taking place at UCLA.

"Good morning," she said as she lowered the microphone. "Thank you for allowing me to return to this summit despite the distressing news already presented. Again, I am going to get right to the point. My team is focusing on the only viable remaining option. That is, reestablishing a normal earth orbit by detonating a chain of nuclear explosions. We have confirmed the size of the massive impulsive force. We must apply an equivalent of 1,613 megatons of TNT in the southern portion of Australia. This force must occur well before the earth is at its maximum position—its perihelion—in our altered solar orbit. Specifically, it must occur within the next two hundred days to ensure conditions here are not too cold for us to properly operate or function. If we wait too long, water flowing through dams will freeze, and we will not be able to generate power to operate machinery and equipment. Information from the research model at UCLA is essential in providing the crucial exact date and time for this event. We expect to receive that information at any moment. Jason Scott remains in close contact with his research lab."

Jason's face appeared on the screen. He seemed dazed, as his mind was adrift with his existing problems. He, too, was unaware of the events that occurred in his lab at that very moment.

Dr. Walker continued. "Obviously, this undertaking is enormous. It requires unprecedented cooperation and trust among the various nations and countries of the world. My team indicated that the level of cooperation and trust needed between world organizations and countries is unprecedented and potentially impossible. However, we have no choice. We must have the support of the UN. I am committed to doing whatever is takes."

"Thank you, Dr. Walker," Secretary-General Loy said. "First, this body has every reason to trust you and your team. Overcoming your personal family challenges is a testament to your commitment and tenacity. You have demonstrated your dedication to NASA and to the people. We must begin somewhere, so we will act on your recommendations."

"Thank you," Dr. Walker replied. "This activity has a beginning, an end, and a specific goal. Thus, we believe it should be organized as a project or program. In other words, it will have a project leader, a timeline of events, and a decision-making and coordination process. For example, construction of the Egyptian pyramids, the Great Wall of China, and putting a man of the moon all had a project structure. The major difference here is the extremely short time frame in which we have to accomplish our goal."

"Feel free to ask for whatever support you need, and we will comply," the secretary-general said.

"We will not hesitate to ask for whatever we need," Dr. Walker concurred. "We have taken the liberty of developing an outline of how we believe the project will be structured. I will give you a brief overview of the plan and then indicate how we must proceed. The challenge we face is enormous. We need real buy-in and world approval to accomplish it within the time constraints we have. On the screen is the outline."

Earth Realignment Project (ERP)
A.   Confirmation of Data and Cause of Phenomena
B.   Verification of Proposed Solution and World Cooperation
C.   Amassing, Assembling, and Deploying Nuclear Bombs
D.   Nuclear Chain Detonation Process
E.   Evacuation of Australia
F.   Constructing the Detonation Silos
G.   Assessment of the Nuclear Aftermath
H.   Protection of World Communities
I.   Rehabilitation of Australia

"Let's focus on the ERP," Dr. Walker said. "One of the first things we must do is confirm and double-check our data. We will thoroughly evaluate and identify all ramifications and consequences of our findings. We must be certain that we are arriving at the correct conclusion based on the mysterious phenomena we have experienced. Unfortunately, we don't have the luxury of time to study and debate our situation. Time is critical. Nonetheless, this confirmation is underway as we speak. Since time is of the essence, we must conduct the verification in parallel with the efforts of the cooperating universities and NASA agencies by initiating the Earth Realignment Project, or ERP. Further, we are working in parallel with the universities and NSAS to check and double-check and determine with absolute certainty that the nuclear detonation approach will work. This is also underway on several fronts.

"A third key parallel activity is to assess all the consequences to our planet and ecosystem from the aftermath of so many nuclear explosions. This activity is also underway on several fronts by the leading authorities in these areas. The preliminary findings indicate that despite serious long-term consequences, we can survive with minimal damage. We will discuss this issue later.

"Before we get into all the details, let's pray we are successful," Dr. Walker said. "I want to add that the team believes we can amass and successfully detonate a phased chain of nuclear explosions. There have been twenty-one thousand nuclear weapons developed around the world over the past years—many of which are now disarmed—each with an explosive force capability ranging from fifteen kilotons to twenty megatons of TNT. The total worldwide yield is about 1,590 megatons, or 1,590,000 kilotons, of TNT. To put things into perspective, the bomb dropped on Hiroshima in 1945 was fifteen kilotons. The current worldwide capacity is thus one hundred thousand times greater. Consequently, we are faced with two major stumbling blocks.

"First, even though we will be using clean nuclear bombs, the radioactivity that would escape into the atmosphere will likely encircle the earth. This radiation is likely to significantly increase the incidence of cancer, birth defects, and other health problems over the next several years unless we find ways to shield ourselves from the radioactive fallout.

"The second concern is more specific and fatal. It is damage to the earth's crust. The detonations could puncture through the outer crust to the core of the earth near the detonation. That would be catastrophic for every human being. We are aggressively developing plans to strategically control and lessen the impact to our planet. If the bomb detonated is far enough belowground in several deep excavated silos, we can achieve the proper blast direction. This is technically similar to firing a sawed-off shotgun. Our calculations indicate we would be dangerously close to rupturing the outer core of the earth in the local region if the blasts are not timed and spread apart with precision. In other words, we could blow a hole into the earth that would destroy all life as we know it.

"This is a synopsis of the realities we are facing. I am being candid to reinforce the good and bad of what is at stake."

"Do we have any other options?" asked Secretary-General Loy.

Dr. Walker was firm. "Unfortunately, there are no other viable options. The enormity of this project is difficult to imagine. One of our team members said it is like building the great pyramids in two hundred days instead of the twenty years it actually took or like sending a man to the moon in not ten years but two hundred days. However, we believe that mankind has the will and the resolve to overcome the greatest threat it has ever faced. Clearly, we don't yet have all the answers, but we believe we can do it. We believe we *must* do it.

"Thank you for your attention. It's now up to you."

Dr. Walker accepted another stack of papers from one of her team members.

"Amassing, assembling, and detonating are only a few of the critical challenges we will face. We will now start focusing on Section E of the ERP," Dr. Walker stated. "This is another critical aspect of the plan and covers evacuation of Australia.

"We have determined it necessary and prudent to evacuate all persons from the continent. This clearly will have a profound impact on the entire earth population. The major challenges include, one, where will thirty-million-plus inhabitants be relocated, and two, what means and modes of transportation will accomplish that. To help you begin contemplating this gargantuan task, we will provide you with some data.

"We believe there are about twenty-seven thousand maritime vessels throughout the world. Even though we lost hundreds of vessels when the mega-tsunami hit the Pacific Rim countries, we can still be successful. It damaged numerous harbors, but we believe we can evacuate all inhabitants of Australia in the time required. We have been meeting with the Australian leadership, and they realize they have no choice but to proceed with ERP. We will need to use all the vacation cruise ships in the world—about 128 ships—and the naval fleets of all countries—972 ships, including forty-nine aircraft carriers. We will also solicit the services of the eleven thousand fishing boats throughout the world. We have a team already working on the evacuation. We will discuss this in more detail later, but for now I need to put forward what we believe is the subtle crux of all these endeavors—Section B. Achieving worldwide cooperation."

Dr. Walker gazed around the room at the sea of bewildered faces.

"Madam Secretary, this aspect will likely be the most difficult and tumultuous aspect of the undertaking. Achieving the worldwide cooperation and trust necessary is more than a mere notion. It requires sincere trust while abandoning old self-defensive policies that have evolved from grudges and prejudices. We believe we have enough technical expertise to be successful, but it will take the collective cooperation and trust from many fronts."

"I can relate to your concerns," the secretary-general said. "We have no choice but to listen and cooperate to the fullest extent possible."

"You're right," Dr. Walker confirmed. "We have never achieved cooperation of this magnitude. One critical first step is to follow the findings of the UN youth commission for achieving world peace. As you know, we formed this commission in July of last year. They have examined current and past world cooperation endeavors and world peace initiatives, including natural disasters, wars, and genocides. They have formulated some profound observations and have successfully deployed and tested them. We gathered these together and formed the guidelines for achieving world peace and harmony. These are success stories, and the key parameters and conditions necessary for those successes have been quantified. Although the team was not scheduled to present their recommendations until next month, it is essential they reveal them to us now so we can use them in this, our most crucial undertaking. As you know, this commission consists of representatives from each country, and they conducted an evaluation of the political, economic, military, and religious systems of each country to formulate their guidelines for peace throughout the world."

Just as the secretary-general was about to introduce the youth commission delegates, one of the security messengers burst into the chamber. Everyone paused, wondering what was going on. The winded security guard held his chest as he relayed his message to the assistant to the secretary-general. The assistant grabbed the microphone without conferring with the secretary-general.

"I am sorry to announce there has been an attack on the UCLA research lab. A radical group of dissidents destroyed the lab and all the personnel within it."

The national representatives within the chamber were stunned and began to grow unruly. Several stood, clenching their fists and waving their hands in the air. Others attempted to leave. The secretary-general motioned to the guard to approach and provide more details.

Jason was horrified at the occurrence of another tragedy. He was so consumed by his own personal tragedies that the people in New York knew of the attack in his research lab before he did. He pulled himself out of his painful mental state and snapped into action. He made a panicked cell phone call to his lab where one of his colleagues explained what had happened. Jason left the room where he was on the teleconference and headed over to his lab.

After he arrived and quickly ascertained that the situation was nearly as grave as announced. The student said Austin was now conscience, talking and taken to the hospital and they had backup data. Jason headed back to the teleconference room and in an attempt to keep the summit going, he tried to downplay the incident and managed to explain the overreaction. He pleaded with the secretary-general to get the attention of the summit representatives and get them seated, but they ignored him and grew increasingly unruly. They began panicking and rushing the doors. Secretary-General Loy and a security guard tried to regain order but to no avail. Frustrated, the secretary-general finally grabbed the rifle from the guard and fired three long rounds into the ceiling. With the microphone next to her, the noise of the rifle burst was amplified throughout the chamber. Everyone ducked to the floor and scrambled under tables.

"Everyone please, please stay where you are!" the secretary-general yelled. "Please be calm. We are overreacting. Return to your seats. Guards, please don't let anyone exit the chamber."

Not realizing what a threatening figure she presented, the secretary-general stood gripping the assault rifle. Everyone began to comply, coming out from under tables and clumsily stumbling back to their seats.

"Please let Jason correct the message," she said.

"Calm down," Jason stated. "Don't let what happened at my lab deter us. I have learned that no one was killed, only injured, and they are all expected to recover, including my best friend. You folks are acting like children. I'm shaken up more than anyone and am heading back to the lab as soon as we finish up here and get my personal back up files of the simulations in the safe."

With Secretary-General Loy still brandishing the assault rifle, no one challenged what Jason had to say. "I agree," the secretary-general stated. "We must control ourselves and not let every obstacle derail our determination. It is imperative that we listen to what the youth commission has to tell us. Their message about needed behavior changes is even more critical during times like this." She started to sit down but realized she was still brandishing the rifle. She immediately handed it to the security guard. "Let's have Dr. Walker continue with the presentation where she left off."

"Thank you," Dr. Walker spoke from her seat. "If we can accept and adopt all or most of the commission's recommendations now without digressing, we might have a chance of achieving success on the ERP project. We believe it is crucial that we depart from evaluating our plight from a technical viewpoint. We need to focus on the main impediments to achieving trust and cooperation. If we behave like we just did, we have absolutely no chance at success."

43

Youth Commission Recommendations

The UN members were restless, victims of the dire situation they faced. The stress was evident on all their faces. They wanted immediate action at this summit.

"Now I would like to introduce Makita Lennox of France and Tanya Calmer of Lima, co-leaders of the youth commission," Dr. Walker announced.

"If they are ready, let's proceed," Secretary-General Loy stated. "Before everyone gets unruly again."

Over the past few years, Makita, Tanya, and their team had earned the trust and confidence of the members of the UN Security Council based on their thorough attention to detail. Makita had propelled herself into the limelight and was respected for her knowledge and understanding of human behavior, especially in the areas of community and regional cultural dynamics. She was a prodigy taking complex issues and simplify them into basic facts and data. Moreover, she had accurately predicted the behavioral outcomes of several crisis events and situations based on the human behavior model she had developed. However, indecisive individuals and organizations frustrated her.

Makita and Tanya walked to the podium and set their documents and laptops carefully on the lectern and table. Makita paused and gazed around the room, noting all the colorful flags draped on the wall of the chamber. She also noted the wide range of dress and regalia worn by the individual members from around the globe. There were multicolored suits, robes, and headpieces. She stood erect, exuding confidence, while she straightened the cuffs on her blue three-piece suit. Standing next to her, Tanya flipped back the delicate sheer veil covering her face and draped it over her floor-length, bright green-and-orange robe. She also stood erect with an air of confidence.

"Good morning, and thank you for inviting us," Makita said in a strong voice. "First, we have some significant recommendations that must be understood and adopted for the Earth Realignment Project to succeed. What we are about to present to you will require open-mindedness and the ability to look forward from where we are today while abandoning some of your past prejudices and principles."

With those words, she looked around at the expectant faces that stood ready to address the challenge that stretched before them.

"Each of you must first do something that is very difficult. You need to release the hatred and distrust of yesterday, no matter how vivid or valid they may be. We are confronted with the most ambitious undertaking man has ever experienced. Most of us, along with the world population, are frightened, confused, and mystified. We would like to believe this is just a nightmare from which we will soon awaken and revert to dealing with our personal problems and challenges of old. But we are not having a bad dream. This is the starkest reality we have ever had to handle. This will affect the lives of billions.

Youth Commission Recommendations

"We believe the project should proceed with the formation of a special organization structure. Even if we can pull off the scientific aspects of the project, the most difficult part, as history has shown, is the human cooperation element. We all recall the difficulties during the Obama administration where the congress was so polarized that many problems never got addressed even though everyone agreed on what those problems were.

"We recommend creating a governing body and appointing a project manager. The project manager would have authority over all aspects of the project and report status, clear roadblocks, and resolve other issues with the governing body. Clearly, for this project to be successful, we cannot have hundreds of separate countries and nations involved in the daily operations and timely decision-making activities. We need to minimize the number of key decision makers by assigning representatives from each continent that carry the authority to represent and speak in a timely manner for all nations within the continent. We must suspend the man-made geological boundaries previously mentioned, such as between the United States and Canada and Israel and Palestine. Thus, we propose two representatives for all of North America, two reps for all of Europe, two reps for all of Africa, and so on, except for the Arctic, which should have only one. This will yield thirteen representatives. These continental leaders need to have full access to sensitive information and be able to make timely decisions on the use and deployment of major resources within the continents' borders. These resources include military personnel, equipment, and supplies."

For the moment, the representatives were anxious about what power and control they must relinquish or partake.

"We welcome you to relinquish passivity, take a deep breath, and focus on the main issues. Now let us direct your attention to the large screen to my right."

Makita stepped back. Tanya stepped forward with her laser pointer and said, "We initially intended that our message would be a documentary. We were preparing to present it at some point in the future. But we have modified it to be of help now, and I will provide commentary and explanations.

"We believe it is worthwhile to focus for a moment on some fundamental behaviors that have worldwide consequences. We have arranged the information according to superpowers. The people and governments of the world's superpowers have a profound responsibility to influence one another to achieve cooperation and trust."

Makita now stepped forward. "As is depicted on the screen, the United States is recognized as the planet's current superpower. Its highest priority is maintaining freedom for its people at nearly any cost.

"For our world civilization to survive, we have less than four months to do the near impossible. Shown on the screen are distinguished individuals performing their very critical jobs of operating the various governments throughout the world, including the powerful senates, congresses, parliaments, the many important state departments, and the Vatican. They each operate on processes, procedures, and protocols established over many years. Unfortunately, if we rely on these well-established institutions, we will be doomed. We must completely revamp our thinking and operate as one extremely efficient world body, and we must accomplish this in four months. We must put control of transportation, heavy equipment, and military manpower under a temporary world-governing body.

"We need to change our behavior starting with the basic family unit. Every household has slightly different rules. One family's rules should not be imposed on another family. Every neighborhood and town consists of many family units and has different values. Every nation and country consists of towns and states that have evolved their own customs. Clearly, the rules, values, and customs of one country should not be imposed on another country."

Makita and Tanya continued speaking on the fundamental philosophies with the world leaders, who appeared to agree with their observations by faint nods. Even leaders from the United States were nodding in tacit agreement.

"To summarize," Tanya said, "our number-one recommendation to the United States is that it must be willing to hand over control of certain national resources needed for mobilization and not impose a litany of constraints. It must back off from imposing its will onto those nations that are hesitant to conform to those of the United States. Adopting this hands-off policy will lead to several policy changes. These are the overarching highlights and recommendations which America's leadership must adopt immediately or our future will be sealed in mistrust and turmoil."

"Time is of the essence," Makita added. "We will now highlight only the most pressing recommendations for the other superpowers. We will conduct sidebar meetings for each nation to allow more thorough review. These discussions will take place immediately after this summit.

"Let's take a look at another emerging superpower involved in international crisis. Israel and Palestine face similar challenges. To ensure egress of equipment and manpower across borders and to the seashores, certain critical changes must be made. The complex relationship and turmoil that exists between Israel and Palestine is not going to be solved during this period. Nevertheless, it is vital that each play an essential role in the ERP project. The two countries must immediately form a commingled, overarching governing body to gain control of critical resources. It should consist of members from both nations and the adjoining countries. This is necessary because each nation of the Middle East has an essential part in deploying manpower, heavy equipment, maritime vessels, and nuclear resources."

As Makita explained in detail what was needed for the EPR to be successful, apparent approval emerged from all sides. A delegate from Israel, standing on the perimeter of the chamber, whispered to his colleagues standing next to him, "These kids have hit the nail right on the head. Can the leaders suspend their differences given the gravity of our situation? Fortunately, I think so."

Makita and Tanya continued, encouraged by the positive reception of their idea.

"Okay, let's talk about another superpower," Makita said.

"A similar situation exists for North and South Korea," Tanya asserted as she stepped forward again. "Similar requests apply as for Israel and Palestine. We know what naval and nuclear resources they possess collectively. These resources are critical for the success of ERP."

She continued to provide compelling arguments to the world leaders. Each presentation was met with polite applause.

Makita and Tanya went on to explain the need for countries to open all man-made borders to allow efficient movement of trains, buses, trucks, and military vehicles. The massive evacuation of Australia required that people be housed. Maritime vessels, including cruise ships, ocean tankers, and yachts, were needed to transport people. They needed resources to abet accumulating nuclear material and the transportation of nuclear devices.

They concluded with acknowledging the UN Security Council for allowing them to present their findings. The audience stood and offered thunderous applause for several minutes. However, in the back of everyone's mind was doubt about achieving the needed cooperation. Many of them feared that if the ERP were not successful, everyone would die and life would cease to exist.

Dr. Braxton listened to the presentation, but he, too, felt it would be impossible to gain the worldwide cooperation necessary, especially considering the time frame. Other countries, such as Portugal, shared his skepticism and did not join in the applause. They knew that every nation would likely want to negotiate every point and argue every issue. Furthermore, they were aware of how many great leaders in the past had tried in vain to achieve just a fraction of what these kids wanted to accomplish in mere days.

Braxton believed the rhetoric and divisiveness between him and the religious community would derail the incredible amount of trust and cooperation needed among governments, politicians, and the religious community. He perceived the religious community as inflexible, unwilling to do anything drastic unless some "sign" was evident and felt they just wanted to stick their heads in the sand and wait. Further, Braxton believed that continuing without general agreement on a well-defined plan of action would be futile.

"Excuse me for interrupting, Madam Secretary," United States Ambassador to the UN Amy Davis interjected. "I have just received a communiqué from the president of the United States. If you will allow me, I will read the statement."

"By all means, proceed," the secretary-general said.

"First, some US citizens will not agree with all of your findings about the United States. Some may agree, and some may totally disagree. But we will not let that detract from implementing your plans. We have a world crisis of epic proportions that we must resolve quickly or none of these lesser issues will have any bearing. Therefore, I have authorized Ambassador Amy Davis to offer the full cooperation of the United States to proceed with your request for resources. We also agree to temporarily relinquish authority of our maritime resources to the control of the continental leaders as requested by the youth commission. We hope this will demonstrate our willingness to make the necessary adjustments in our behavior during this interim crisis period."

Surprised, Secretary-General Loy thought, The United States offering such unconditional support is incredible and encouraging.

"You also have my country's full support," the prime minister from North Korea offered. "We will adopt and abide by the recommendations from the youth commission."

Youth Commission Recommendations

Several other countries came forward with similar assertions, including Israel and South Korea.

"I am flabbergasted," said Secretary-General Loy. "I am going to abandon normal protocol and proceed with my gut feeling. Let's take a quick vote to ascertain whether to adopt the specific recommendations from the youth commission report even though we have not studied all the details. All those in favor, please vote now."

Several hands went up.

"All those opposed?"

No hands went up.

Astounded, the secretary-general took a deep breath and cleared her throat. "I guess there are no countries opposed to adopting the recommendations 'as is.' Well, then, it appears there is a glimmer of hope for us after all."

Dr. Braxton frowned, unable to believe or comprehend what he had just witnessed.

But all was not well in the chamber as one country had abstained from voting. The frustrated member from Portugal who lost his family from the Tsunami was overtaken by anxiety and frustration, jumped out of his seat and charged the front table. "This is all rubbish! You are wasting our time with this 'trust and cooperation' stuff. We are all going to die!" He grabbed a heavy wooden placard and hurled it at Tanya, striking her in the face and inflicting a vicious gash on her cheek. Several people tackled him and wrestled him to the floor.

"You are doing the wrong thing!" he screamed. "You must stop the destruction of our earth! Don't let them convince you to blow up our fathers' planet. Stop them now!"

During the melee, Tanya slumped over the table, bleeding profusely. Makita, although terrified, was furious with the behavior of the member as she helped her colleague. The medical staff also rushed to help Tanya while the security guards began to drag the member from Portugal to the security office.

Professor Stein was watching from his seat at the summit, and his mind snapped. Hobbling over to the member from Portugal, he delivered a solid blow to the back of his head with his cane and continued pummeling him until he fell to the floor. The surprised security guards struggled to restrain the professor and managed to pull him away before he killed the other member.

The secretary-general finally called a halt to the entire proceedings.

44

Tense Summit

The summit chamber was now in shambles. There was chaos outside and turmoil inside. The members obliterated the fundamental concept of cooperative behavior and trust before it started. Within minutes after agreeing to a plan, they were at one another's throats and things were already derailed. It appeared virtually impossible for the nations to achieve the trust and cooperation essential for mobilizing the equipment, materials, and personnel necessary to execute the massive ERP project.

After a short break, Makita returned to the chambers and marched up to the microphone on the front table. "Tanya gave me a message," she announced as tears streamed down her face. "I just left her. They think she will be all right. She will need stitches, and there will likely be some permanent scaring. But she insisted that I continue.

"One of the things Tanya and I had intended to emphasize in the presentation was for us to focus on the things we have immediate control over. The past is history, but we can affect the future. None of us has control over the unknown. But we can impact the future by creating plans that will save lives. We must stay the course and develop and implement the ERP plan step by step to control the fate of humanity."

Many of the members were touched by the comments from such a courageous and mature young person.

"Let's proceed with the remainder of the presentation," Secretary-General Loy announced with stern conviction. "We know what we have to do. It appears to me that we cannot give up now because of the actions of one person or even a few. The vote indicated that everyone is unified with the same sense of urgency that appears to be transcending cultural boundaries. Dr. Walker, please continue where you left off."

"Thank you," Dr. Walker said. "Then let the vote stand, and we will assume that the adoption of the new resource mobilization policy changes will be implemented. Let's focus on what else we must do."

As Dr. Walker continued with further details of the roles of the continental leaders, Ambassador Amy Davis spoke. "Excuse me, Madam, but I just received word from the president again. She has consented to invoke her executive privilege as commander-in-chief and temporarily provide key military resources to these newly selected continental leaders."

It became immediately clear to all the member nations that this unconditional trust demonstrated tremendous courage never before exhibited by any superpower of the past. No military leader would ever dream of making such a bold gesture. The chamber erupted with applause for the courageous action taken by the US president. This action set the stage for future barriers to tumble.

"Unbelievable," Dr. Walker exclaimed. "Such trust and cooperation we only dreamed possible. I am surprised and very pleased. We will provide more details of these activities and the specific selection process for the continental leaders later in the presentation."

Tense Summit

Secretary-General Loy gave Dr. Walker a thumbs-up and then stood up slowly and raised her hands above her head in applause, holding her fingers in a "V." Ambassador Davis also began clapping and slowly stood up. Then the representative from Canada joined in and was soon followed by an avalanche of clapping throughout the room. As they stood, announcements from the leaders of other superpower nations began pouring in, committing to relinquishing their power during the ERP project.

Cardinal Malloy and all the religious leaders stood up and joined in the applause. However, one skeptical leader leaned over to one of the cardinal's colleagues and said, "If man can simply move the earth, why do we need God?"

"Nonetheless," the happy religious leader responded to his neighbor, "be happy the scientific and religious communities have finally come together, and praise God for that! We will move the earth if it is God's will."

The show of support and solidarity spread throughout the chamber and in every country that was watching the event unfold.

"This is a miracle," Secretary-General Loy said. She could no longer contain her emotions, and her watery eyes glimmered and her heart pounded with joy. "Only time will tell if the project is successful," she stated. "But right now we are experiencing a miracle. Every nation and country is supporting the project. I guess all the religious vigils paid off. This is the proudest moment of my life, to witness so many diverse religions, cultures, and races coming together. No matter what happens to me, at this moment I can say I am proud to be a part of the human race."

Euphoria radiated from her eyes as well as from those of her colleagues. Professor Stein looked at Jason on the screen and gave him a thumbs-up.

The affirmed atheist Dr. Braxton, filled with emotion, began perspiring and trembling. He bowed his head, brought his hands together, and said to himself, "Thank you, God. I believe. Hallelujah," and the many years of doubt and disbelief began to fade. There was a God and an almighty spirit present after all. He clasped his trembling hands together so tightly that his fingers grew numb and his arms began twitching. Something greater than anything he had ever imagined had to be present for such a miracle to occur.

45

Devastation at Summit

Dr. Braxton peered at the table, emotionally drained. With his head bowed, he stared at a half-empty glass of water. Suddenly, tiny, vibrating ripples appeared on the water's surface. In the background, he heard a strange, low-level whistling. The water ripples began to enlarge, and the whistling escalated into an eerie groan.

The entire table began to vibrate with a distinct hum. Braxton carefully stood up and looked around the room, hoping to extract a clue about these strange disturbances. The distant groan was increasing.

Dr. Strickland looked as puzzled as Dr. Braxton. Secretary-General Loy stood up and tightly gripped her chair, looking around the chamber. "What's happening?" she asked.

Sonia Walker froze in her seat while Makita and the others at the center table slid under the table. The walls began to buzz, and the ceiling rattled. Several of the UN delegates, now terrified, drifted toward the chamber exit.

The huge chandeliers began to sway and debris trickled from the ceiling. The groaning sounded like a freight train grinding down the tracks. The atmospheric pressure in the room abruptly increased. Papers swirled toward the ceiling while books flew to the floor.

Suddenly realizing what was happening, Dr. Braxton yelled, "Tornado!"

The walls rattled violently and bulged. The large, ornate glass-dome-ceilinged section of the chamber suddenly shattered, spraying shards of glass everywhere. Papers, books, and chairs flipped over, swirled around, and became airborne as air surged through the ceiling.

Part of the funnel engulfed the building, abruptly sucking tiny Makita from her chair and ejecting her through the shattered dome ceiling. Several UN delegates became airborne along with the swirling debris and were catapulted behind Tanya through the open ceiling directly into the eye of the tornado.

The secretary-general and a few others desperately clung to a railing attached to the floor. The south wall of the chamber cracked and shattered, causing an explosion and propelling debris in all directions. The tornado then sucked several more through the open wall and skyward to their fate.

Most of those in the room hunkered down for dear life between a pillar and a rail. Through the gaping hole in the wall, they saw billboards, food tables, chairs, and people swirling around like matchsticks. Farther down the street, they saw three funnel clouds ripping apart buildings along the Manhattan financial district. As the barrage of tornadoes faded in the distance, debris was strewn everywhere and carnage left in its wake.

Dr. Braxton, frozen with shock, clenched the railing, too petrified to budge. But the gut- wrenching groans, moans, and screams around him compelled him to help the others. Next to him laid Dr. Strickland, who had been impaled by flying debris. Adjacent to the center table laid several UN delegates, crushed and mangled by the falling ceiling.

Devastation at Summit

Dr. Braxton and Secretary-General Loy searched for survivors. Sonia Walker, bruised and covered with brick debris, was otherwise okay. Dr. Braxton cringed when he found Professor Stein's contorted body that had been crushed by a fallen ceiling beam. Jason also saw Stein's crushed body and collapsed to his knees in anguish.

46

Fatal Execution Flaw

With the demise of many key members of the UN summit, it was difficult to go on, but the survivors of the tornadoes in Manhattan and other places around the world became even more focused and motivated to continue with the only viable option presented at the summit. Fortunately, thoroughly documented plans allowed others to step up to the plate and take over for their deceased colleagues.

It was clear that the logistics involved in attempting to vacate an entire continent by mobilizing every seaworthy vessel in the world under the constant threat of environmental mayhem was impossible. The resources required, not to mention the level of cooperation, was a pipe dream—foolish, impractical, and, frankly, unachievable. Such was the belief of many naysayers.

The sporadic tornadoes and lethal lightning were wreaking havoc around the world. The mega-tsunami had not fully subsided, and the full death toll was not yet known. The public was unaware that many of the governments around the world believed the ERP project was doomed from the beginning. Further, escalating public fear inspired radical groups to infiltrate the Respect Mother Nature Society, as each wanted to implement its plan to undermine the planned ERP detonation.

Fatal Execution Flaw

Nonetheless, as the days, weeks, and months passed, the ERP world community gained momentum and began to rally around the project while suspending almost all normal activities. Most people had restricted their movement outdoors in fear of the unpredictable lightning and tornadoes. The ocean swell had sloshed to an enormous scale, wreaking havoc on ocean vessels and ports around the Pacific Rim countries. Fortunately, the ERP personnel had developed an effective and robust worldwide communication network using the Internet, news, and social media. Websites proliferated to help identify every conceivable aspect of the project.

Government personnel and private citizens from every country had become aligned and cooperative with the project, and every country with nuclear capability had begun implementing its plan to disarm its nuclear weapons and reconfigured them to conform to the detonation requirements before shipping them to Australia to become an integral part of the explosion process. Each country had to implement extreme precautionary measures to protect against the sporadic lightning. Iran revealed it had secretly developed four high-grade nuclear weapons in underground cities but were now dismantling them for use at the detonation zone in Australia. Iraq disclosed it also had two weapons of mass destruction hidden in another underground location. Israel revealed it had an arsenal of several hundred large nuclear weapons for use in the ERP.

This was great news for nearly everyone. However, it was rumored that when former American president Bush heard Iraq's surprising declaration he had to be restrained due to the nefarious "weapons of mass destruction" issue that caused havoc during his second term in office. Nonetheless, each country continued to support the ERP in a spirit of trust and cooperation.

Ninety percent of the world's military sea vessels were reconfigured to accommodate the designated passenger load venturing from Australia to the planned destination countries. The comprehensive plan called for smaller vessels to deliver the passengers to countries closest to Australia while the larger vessels, such as cruise ships, would transport their passengers to the most distant countries.

As one UN-led fleet of international ships approached the harbor in Australia, the newly formed continental leaders met their first challenge. The fleet was met by a barricade of dissident Australians who had organized a human blockade in a key dock in southern Australia, which was successful in preventing the ships from entering the harbors to begin the evacuation. The ERP security headquarters did not intervene with military might but instead allowed the continental leaders to work together to resolve their first major crisis. With patience, international diplomacy, and prayer, the continental leaders eventually obtained buy-in from the dissidents. Videos and graphics revealed to the dissidents why the ERP was necessary. So persuasive was the evidence at hand, it wound up converting the dissidents into staunch advocates who then allowed the evacuation process to proceed.

Most local and worldwide charitable organizations were involved in food distribution to the passengers while en route to their new destinations. An effective and well-coordinated family housing effort for the displaced Australians had emerged, as families from all over the world opened their doors to accommodate them.

Fatal Execution Flaw

New church members packed into stadiums, churches, synagogues, temples, sanctuaries, mosques, and other religious structures. The number of people claiming to have found God had skyrocketed to the point that institutions had to conduct special prayer services. Religious material in most stores had sold out. However, many of the religious facilities and buildings were converted throughout the world into housing for the Australian immigrants.

As the ERP project progressed, every scientist on the team was intensely focused on his or her job. However, in the back of their minds they worried about the possibility of an omission or a miscalculation, as having overlooked a fatal scenario would be demoralizing. As fate would have it, their greatest fears were about to materialize.

A student at Oxford doing research in archaeology inadvertently uncovered something that stood to terminate the entire ERP effort. The student was documenting the generally accepted hypothesis that most if not all of the dinosaurs became extinct over sixty-five million years ago when a gigantic meteor struck the earth, sending dust and a plume of debris into the atmosphere that eventually blanketed the earth and subsequently wiped out much of the earth's vegetation. The dinosaurs that survived the blast eventually starved to death. Based on his calculations of the size of the impact crater, the student determined this meteor struck the earth with a force equivalent to forty megatons of TNT. He knew that the size of the nuclear detonation planned for the ERP project in Australia was nearly sixteen hundred megatons of TNT—a force forty times greater than the force that had killed the dinosaurs.

The student knew he must immediately inform his adviser so they could contact the right people to suspend all efforts. He and is advisor quickly forwarded his findings to the chancellor of Oxford University, who in turn contacted the ERP public relations office.

Shaken by the findings, the chancellor arranged an emergency teleconference between the Oxford University research department, the ERP project manager and chief scientist, and the continental leaders. The Oxford student began explaining the dinosaurs' extinction research and the calculations of the meteor impact. The ERP chief scientist abruptly interrupted the discussion. He had been one of the key drivers and decision-makers on the project and was now highly regarded worldwide.

"Hold on a moment," he said to the student. "Before you go any further and get everyone bent out of shape, let me say something. First, we are not going to detonate a sixteen-hundred-megaton bomb. Our maximum detonation will be twenty megatons. The other detonations will be smaller and staggered over a short time period and location, albeit in seconds, but not instantly or simultaneously. Second, each of these detonations will be placed in deep silos or wells encased in thousands of feet of ablative concrete to create what you might call a sawed-off shotgun effect, giving rise to a big kick, or jolt in our case. I hate the word *explosion*. We have had breakthrough technology for some time now. We now have controlled nuclear energy release, or CNER. We designed each CNER to create a nuclear clean, pencil-like beam of blast force rather than an explosion that spreads inefficiently in a lateral direction on the ground like a mushroom cloud from a meteor impact—the type of explosion that killed the dinosaurs. We believe the earth can withstand multiple staggered and separate CNER forces of several thousand megatons dispersed over thousands of miles along the large detonation zone in Australia. This will not destroy all life on Earth. In fact, we hope no human life will be lost at all."

He paused for a moment, the room filling with silence.

"Nonetheless, our major concern is rupturing the earth's outer crust," he continued. "The earth is a layered planet consisting of an outer crust over eight miles thick, a mantle, an outer core, and an inner core. One of the initial indicators of the survival of our planet would be the occurrence of a nuclear blast plume that is either red or brown. What would instantly signal disaster would be a bright-red continuous plume. Pray that we do not see this because it would indicate that we have punctured through to the earth's outer core. The molten magma would gush more ferocious than all the combined volcanoes in man's history. This would destroy all life, as the momentum surge would spread over the entire planet. However, a brown or gray plume would indicate the detonation was successful, and that the detonation force had helped propel the earth back into the right orbit.

"My point here is that we are not designing the detonations to occur at exactly the same time and location. Again, we emphasize that they will be staggered, creating a sustained, impulsive force. In some ways, it is analogous to slowly letting air out of a balloon. By the way, this is exactly how we currently handle transfer orbits of satellites. To get any satellite into a higher or different orbit around the earth, we must use thrusters to apply impulsive force for a specified time—I am speaking of thrust impulse. In our case, the satellite is the earth itself orbiting around the sun. We will sustain the nuclear detonation forces over several minutes, transferring the earth back into its original, near-circular orbit rather than its current highly elliptical orbit. Does all this rambling explanation help you?"

"Well, well, chief," sighed the public relations manager. "That explanation is a relief to me. Do you Oxford blokes buy into that explanation or do you have additional concerns?"

"No," the research student said. "The chief addressed the issue I was concerned about. It sounds like the ERP project team is on top of things."

"I want to thank you for bringing up this concern, and don't hesitate to question things in the future," said the public relations manager. "We need everyone to remain vigilant to ensure we don't make any catastrophic mistakes. Thanks again." As the teleconference concluded, he thought, I hope another potentially catastrophic finding doesn't surface. The next time there may not be any answer.

When the news of the averted showstopper finally reached Dr. Braxton, it touched the newly religious spirit deep within him, triggering a radical philosophical issue he was about to face. He became seriously concerned, and thoughts of abandoning the ERP project persisted. He could not reconcile allowing his beloved Australia, with all his relatives, to be destroyed. "Let nature take her course," continuously echoed in his mind. He approached President Clemson and the continental leaders to consider abandoning the ERP nuclear detonation effort.

Even so, as the days passed, Dr. Braxton's efforts proved futile. The momentum for EPR was firmly entrenched and well underway. When the leadership ignored him, he became enraged and began sympathizing with the doomsday radicals organized by Professor Thatcher and his team. Fear and anxiety was taking its toll on the few individuals who thought the entire ERP effort was in conflict with the very essence of nature.

When Professor Thatcher became aware of Dr. Braxton's reversed views, he worked to recruit him and succeeded. Their plan was to release chemical weapons on the ERP fieldworkers just before the nuclear detonation. And with Dr. Braxton's help they just might succeed.

47

The Hospital Visit

Several months after the start of the project, Austin was still recovering from the injuries he suffered in the attack at the lab. UCLA and the United States government had worked around the clock and managed to recover much of the data using Jason's backup files from his safe and were able to reconfigure the simulation model with newer, faster computers. The lab had been reestablished on campus, but at an undisclosed location with plainclothes security people monitoring the operation.

Jason and the government officials continued their search for Jane. Jason was unrelenting in his pursuit and remained undaunted by the fear that gripped the world. He cross-examined Jane's parents, grilled her friends and colleagues on campus, and staked out her apartment complex. No clues, nothing. Now exhausted and dejected, he was on the brink of collapse.

During this time, Austin stopped being so self-centered, as he had been focusing on himself and his own health. He had not heard from Jason lately but knew he was continuing his relentless search for Jane.

As a condition of her release, Jason's sister Britney had to be committed to the eighth-floor psychiatric ward of the hospital in Harbor City, California, for evaluation and counseling. Austin decided to visit her. He made his way to the hospital, observing many abandoned scorched out cars from lightning strikes, ripped up trees from tornadoes, desolate streets, and the few people he did see wore multilayer clothing and thick coats to brave the chilling conditions.

Austin went up the elevator and sluggishly approached Britney's room, as his neck and back brace protected his injuries but restricted his movement. He peered in the door and noticed several posters of various rock groups and lead singers plastered on the wall. Britney was delighted to have a visitor, especially Austin. Immediately, she became energized and bubbly.

"What up, Britney?" Austin said as Britney walked over to give him an awkward hug before returning to bed. He glanced at the many rings on her fingers.

"How the hell are you, Austin? You look like a mess."

"I'm okay. I'll be out of this contraption soon. How are you?"

"I'm fine. I hate being incarcerated. But I guess I would rather be in here than outside dodging lightning bolts and tornadoes and worrying if they are going to blow up the earth. You look awfully uncomfortable in that contraption, but I'm happy to see you anyway. How the heck are you?"

Austin eased down in the chair next to her bed. "Not too bad. I don't have any significant pain from the injury. They say I will fully recover. I just have to wear this contraption for a while."

"I understand they still haven't caught the people who did this to you."

"I hope they don't catch them."

"What? Why not?"

"I want to catch them myself."

"Austin, come on. What would you do?"

"Have you ever seen a man with his feet tied together, hands tied behind his head, and hanging from a tree by his testicles and penis?"

"Austin, come on," Britney said with a snicker.

"Just wait."

"Fine," Britney said. "Anyway, thanks for coming to see me. Maybe you can tell me what's happening out there because Jason doesn't say much when he visits. He is really devastated, and it shows."

"I know. He has been avoiding everyone. How much do you know about what's going on with the ERP activity?"

"I don't want to hear about that stuff. I've heard enough from Jason, even though it's been awhile. Tell me what's happening with you and your friends?"

Austin fished around to determine how much Jason had told Britney about what had happened between him and Jane. "Has Manuela been to see you?" he asked.

"No, she called once and said her broken foot from the campus tumble was healing, but I haven't seen her in a while. Have you?"

"I think she's gone into seclusion, waiting for Jane to return. She feels guilty and blames herself for her leaving."

"You know, Jason won't talk about Jane at all when he visits," Britney said, the sadness showing on her face.

"I don't know anything about her whereabouts either. We just hope she's still alive."

"Jason probably spends all of his time at the lab with that cocky chairman and that professor," Britney said.

"What?" Austin said, puzzled. "Didn't Jason tell you about Chairman Chapman?

"No. What about him?"

Austin leaned back in his chair with his hands holding onto his back brace. "Well, well, well," he said. "Crew-cut Chapman was forced to resign by the board of directors and the university chancellor. They were outraged at him for succumbing to piglet Thatcher's demands. They believed he sent a letter to the Nobel Prize committee reprimanding Jason for the tumbling incident. After Chapman's dismissal, they planned to send Professor Thatcher back to Harvard before his work was finished here, but Harvard refused to take him. Thatcher's prior conflicts with students and his recent run-ins at UCLA finally caught up with him. He is out on the street now. In fact, I think the authorities are looking for him."

"Whoa!" Britney exclaimed. "I think they both got what they deserved. What goes around comes around. Maybe the same thing will happen to the NASA jerks for sending my parents on that ridiculous experiment. Too bad I wasn't able to give them what they deserve and finish them off."

"Hey, let's not go there. That's why the authorities sent you here. Let me tell you some good news."

"What?"

"Tyresha and I are engaged."

"You're kidding. I knew you were seeing her, but engaged? Oh my God! Congratulations!"

"Yes. I have dropped all my other girlfriends just for her."

"That is so cool, Austin. I always thought you guys were right for each other."

"Thanks."

"I am so glad you have put aside chasing other women. Too bad some of the other guys don't follow your lead. Like Luther."

Austin looked puzzled again, and the smile fell from his face. "You haven't heard about Luther either?"

"No. What about him?"

"He has AIDS, is very sick, and is undergoing treatment."

"How awful!"

"His friend told me some necessary unusual treatment might make him impotent."

"Oh my God, can you imagine how he must feel, being such a ladies man and all? I guess he needs to focus on just trying to live now."

"Yeah, it's tragic. Well, I have to go. I'll come back soon. Feel free to call me anytime."

"I will, just as soon as these jerks let me make outside calls. I'll see you later, Austin. Thanks for dropping by."

Britney would have come unglued if she knew that Jason had temporarily transferred advisory authority of Britney to Austin due to Jason's and his parents' absence. In fact, Austin was there to help the authorities determine Britney's release from the psychiatric ward. Unfortunately, he had to tell them that she continued to harbor strong resentment toward NASA.

48

Realignment Project Hardships

Several months after the inception of the ERP, people the world over remained tense. Since the earth was in a new orbit and moving farther away from the sun than ever before, profound environmental changes continued to escalate. The temperatures throughout every region in the world began to plummet. The passenger transportation across the oceans and seas was growing more treacherous due to increased ice formations. The public was now keenly aware that conditions on the planet had deteriorated to such an extent that the effort was at risk.

The ERP leadership invoked special precautions on the use of tools and equipment in the transportation, handling, and assembly of the nuclear bombs and detonation devices. Because of the cold temperatures, the detonation team members on site in Australia had to wear specially insulated gloves that were cumbersome when handling the delicate tools and nuclear devices.

One of the three-person teams conducting the delicate device assembly process had been working outside for several hours. The captain saw that they had been unproductive over the past few hours and grew irate at their lack of progress. He stormed out of his tent to confront the team and reprimand them for not responding to the urgency of the situation.

The team gave him no respect and ignored him for the most part, which made him even more furious.

When he made his way behind them, they didn't even have the courtesy to turn around and face him. He stormed around in front of them, but they continued to ignore him. Then the captain noticed that they were motionless, and that his lead man had ice over his eyelids. When the captain looked closer, he believed he detected complete disobedience. Shouting, the captain snatched the lead man's hand. The hand snapped off and fell to the ground. The captain jumped back in horror and radioed his experience to headquarters.

The continental leaders led by Secretary-General Loy were devastated. Dr. Braxton realized that some of the workers who had perished in the field were his relatives and family members.

"My God, what is happening?" We must all help these heroic people." While waiting for the continental leaders to respond, he was heartbroken over the loss of his homeland and some of his relatives. He then thought about Jason and how he, too, had really suffered from tragedy and hardship with the brutal attack of his lab colleague and best friend, Austin; the death of his professor, Stein; the incarceration of his sister, Britney; the loss of his parents; and the loss of his fiancée, Jane.

Braxton felt compelled to contact Jason. After an extensive search using government officials, he was able to contact him by phone.

"Jason, we share some similar tragedies, and I offer you my condolences," he stated. "I have experienced personal loss, but you have also lost much, including Professor Stein and your parents and fiancée."

"Thanks for calling, but you only need express condolences for Professor Stein. The others are fine," Jason said. "We just need to find them."

"Okay, well, sooner or later, I hope you accept reality."

The continental leaders realized that the critical deadline was approaching. Everything should have been in place by now for the nuclear detonation chain, but the deteriorating conditions had forestalled progress. It soon became clear that they would have to do the unthinkable and detonate the explosion several weeks early even though Aussie people were still being evacuated. Furthermore, there would still be people on the continent.

The ERP continental leaders led by Secretary-General Loy convened an impromptu emergency meeting to deal with the crisis. They invited some key individuals, including Dr. Braxton, Cardinal Malloy, the ERP technical director, and, not surprisingly, Jason via teleconference to reconfirm the earth orbit predicament. The continental leaders had total authority to make the final decision on all ERP issues, as they rule by consensus rather than by majority. However, they wanted to hear any last-minute concerns before proceeding with the earlier-than-planned detonation. Based on the earth's rapidly deteriorating conditions, everyone was in agreement except Cardinal Malloy and Dr. Braxton.

"It appears to me the risk of proceeding early with the detonation is too risky," Dr. Braxton said. "It's likely to seal the fate of mankind rather than allowing nature to take its course. I believe nature has done reasonably well during the past five billion years."

None of those present were aware of Dr. Braxton's recruitment by the Respect Mother Nature Society, which he had secretly provided with information so they could thwart the detonation.

"I agree," the Cardinal said. "The will of the Lord will be done, and He will protect us."

"Does anyone else have any other comments?" The secretary-general said as she looked around the room.

No one responded.

"What about you, Jason?" she asked.

"Whatever makes sense to all of you is fine by me," Jason said. "I'm focused on my parents, sister, and fiancée. I will have my team run another simulation to determine if any tweaks to the detonation impulse are necessary and relay that information back to the ERP team, but I don't think they will recommend any changes. Other than that, I will be focusing on my personal problems."

The ERP leaders decided they had no choice but to detonate four weeks early, as the opportunity for success could vanish at any time. Many argued they needed more time to complete the evacuation. But they had run out of time.

As the countdown began, the ERP leadership informed the public of the explosion that would jolt the earth and lead to mammoth earthquakes in many regions as well as mega-tsunamis. The lives of those in mid-evacuation would be in extreme jeopardy.

Those stranded in Australia were ushered into distant underground caverns in the hope that they would survive the engineered nuclear event.

The wildlife authorities of Australia still quarreled with the engineers and scientists. They wanted to stop them from setting off a detonation that would surely annihilate all wildlife in Australia, which was unimaginable to the wildlife authorities. They demanded that many more resources go toward animal rescue and preservation than was currently expended while continuing to usher as many animals of a given species as humanly possible into their makeshift shelters. They had built the shelters in a valley refuge away from the detonation zone that was surrounded by mountains on both sides. Some reporters had labeled their effort a twentieth- century Noah's ark. Undaunted, they continued their protest and rescue efforts right to the bitter end.

The questions remained "Will all this be in vain, and will all life as we know it be destroyed?" Either the nuclear effects would rupture the earth or the earth would become a frozen wasteland due to its uncorrected orbit.

49

The Unexpected Phone Call

Jason passed the instruction from the ERP leadership to his research team to run the last simulation for the detonation force needed for realignment four weeks early and asked them to forward any revisions directly to the ERP team.

Sitting numb in front of his computer screen, Jason was startled when his private cell phone rang for the first time in a long time. He nervously picked it up.

"Jason, it's me. Jane," a distant, weak voice said.

Jason's heart pounded, and his hands trembled. He was so stunned that he could not say a word. Was he actually hearing the voice of his true love?

"Jason, I've moved. I am now in a shelter house. I have been thinking about you every moment of every day, and I simply had to call you."

Jason remained speechless, unable to utter any words to express his feelings.

Jane's fading voice was weak and apologetic. "Jason, I didn't know what to do. I finally decided I must have an abortion. I don't think I'm going to make it. There is no one here with me now. I feel so weak . . . so weak . . ."

"Jane!" Jason shouted. "Everything will be okay. I'll be right there. Where are you? Jane, Jane, say something! Jane, *please* say something! You can't go. You don't understand. The baby is *ours*, Jane. I love you. Please don't go."

Jason's knees buckled, and he fell to the floor in a fetal position. He felt from Jane's muddled, wheezing voice that she was dying. He lay there trembling and traumatized. His true, divine love was gone, and their unborn child was about to be snatched away before his life ever began. Jane had called him in desperation, and he had failed her. He would never see her beautiful smile again, hear her soft voice, or touch her tender body.

His mind flashed back to when he first saw the elegant Jane facilitating her debate team. He recalled his disturbing concerns over her departure to Oaxaca and the utter joy in her eyes when he dropped to his knees and proposed to her on the airstrip. He remembered the devotion she had provided him on the perilous flight to Vandenberg and his outburst when he found out the baby was actually his. But now everyone was gone; both of his parents, his devoted Professor Stein, and his beloved Jane. They could no longer shelter him from the harsh reality of life's immense, unbearable pain.

Jason stopped feeling sorry for himself and checked the phone number Jane had called from. It was indeed from a women's shelter in Los Angeles. After finding its address online, he dashed to his car and raced to the shelter. The freeways and streets were cluttered with debris due to lack of maintenance since all resources had been placed elsewhere since the onset of the ERP.

When he finally arrived at the shelter, it was abandoned. Jason climbed over the fence and peered through the window. Although it was dark, he detected a dim light down the hall. He broke the window with a rock, climbed inside, and ran toward the light.

When Jason entered the room, he saw his beloved Jane lying on the bed, weak, dehydrated, and cold.

"Jane, it's me! I'm here!" Jason ran to her and embraced her.

She opened her eyes in disbelief and smiled. "Jason, is it really you?" she murmured.

"Yes, I'm really here. Everything is going to be alright now. *I'm* the baby's father. It is *our* child. They ran some paternity test at the UCLA lab and confirmed I'm the father."

"What are you talking about?"

"It's true Jane. I went to the lab and Dr. Lawson compared my DNA with the baby's after you went in for the pregnancy test. You were oblivious but Luther never raped you. He secretly confessed to my friend that he didn't even fondle you. It's our baby."

"You mean my baby and I am starving to death for no reason."

They remained locked in a deep, loving embrace. He took her to the sparsely supported hospital for an emergency check-up and IV hydration. After the brief treatment they decided to head back to the lab to join the rest of the UCLA team for the final countdown of the detonation. After they arrived Jason had arranged for Manuela to drive his discharged sister Britney to the UCLA research lab basement where they all could be together.

50

D-Day

The countdown commenced. A sequence of warning horns and sirens sounded throughout the detonation site, and an enormous nuclear detonation chain commenced.

People felt the shock throughout the entire Australian continent, sending many not hunkered down to the ground with a whiplash motion. Structures began to sway, and some began to collapse. The combined jolt measured 9.5 on the Richter scale. Most people had hunkered down in their shelters, huddling with loved ones in anticipation of the shock. Pets panicked, and birds scattered in flight while wild animals stampeded. Communication lines were knocked out. A wave rippled across Australia and across the oceans, creating mega-tidal waves, and coastal cities the world over braced for the deluge. The propagating ice latent ocean water forged a path that threatened to submerge many of the smaller islands.

From a distant island a few miles off the coast of Australia, the ERP surveillance team looked on as everyone within visible range waited to see if the detonation would result in the end of humanity and all life on Earth. Then their worst nightmare began to unfold.

A fiery, red-and-yellow, funnel-shaped plume, resembling the thrust of gases from a spacecraft launch turned upside down, shot straight up into the atmosphere. It appeared that the scientists had punctured a gigantic hole to the core of the earth. If so, deep cracks and fissures would emerge, and in a short while, the molten guts of the earth would spew out onto the surface of the planet, annihilating everything in its path.

Dr. Braxton, Secretary-General Loy, and others positioned on the monitoring site of the island witnessed this horrific scene firsthand. Dr. Braxton dropped to his knees, imagining the imminent demise of every human being, first in Australia and then on the entire planet. The ERP team had failed to listen to him and to Mother Nature, and no one, including the Creator, was going to pull them out of this one. This is the end, he thought.

Taunt cheeks and quivering mouths appeared on the faces of some of the onlookers while others simply trembled in fear. The entire gory event appeared on TV for the whole world to see just before the electrical power began to fail, taking TV communication with it. People huddled in their homes with family members and in churches, synagogues, temples, and stadiums. They cringed in fear and braced themselves for the unprecedented and tumultuous geological upheaval. Their reactions ran the gamut as, during the last transmissions of the news, reporters vainly attempted to put a positive spin on the numbing events. Communication would soon be lost to the rest of the world as the shock wave from the massive detonation reverberated across land and sea.

Amid all of the gloom, expert volcanic geoscientists, hunkered down at the monitoring island site, stared in awe at the streaming plume that began to engulf them. The plume gradually began to transform from a bright yellow glow into something with streaks of brown, gray, and black.

One of the geoscientists yelled, "Ruptured magma chamber!"

His colleagues jumped up, exclaiming, "It only punctured a *local* magma chamber—not the crust! It worked! It worked! We didn't puncture though to the earth's core! Look at the exhaust plume. It's brown and black. It's humongous! It's beautiful! It's working!"

They danced around the room, hugging and high-fiving. The rest of the on-site scientific team dropped to their knees, sighing with relief. And as civilians began to realize that their initial observations were wrong, they joined in the celebration.

Near hysteria quickly spread throughout the region. The massive detonation had yet to disrupt a few communications lines, allowing jubilant news reporters to sneak across the remaining airways.

But despite the apparent good news, NASA had no way of knowing if the sequenced detonation impulse had jarred the earth back into position. Key communication lines were interrupted. Two questions remained: "Was this monumental effort a failure? Is our planet still doomed?"

51

The Final Outcome

Awaiting the onslaught of devastation, the NASA Command and Control Center team in Virginia continued scanning the skies.

From his control console, one scientist began picking up a faint, distant signal from outer space. As he zeroed in and amplified it, he realized it was coming from an unknown space vehicle. As he continued to hone in on the faint signal, he began to hear what he believed to be voices. From the communication code signature, he realized it had to be coming from within the lost International Space Station. He heard what sounded like the voice of the commander. But believing the ISS had been lost in space, he dismissed the signal and the voices as evidence of his fatigue. He nonchalantly moved the scanner to a different frequency and continued to complete his tasks.

An intern happened to be listening in on the same communication and, in his open-minded eagerness, did not come to the same conclusion. He brought the signal to the attention of his coworker, and they began arguing over its authenticity. Finally, they agreed it might be worth a second look. The intern dashed out of the room to alert the module chief of his suspicion. The chief, along with several of the support staff, patched into the communication network line and eventually picked up the faint signal. The chief was astonished that he might actually be in contact with the space station believed to be lost forever.

"Come in! Please, come in!" the chief said. "Please acknowledge. Please identify yourself."

"This is Dorothy Scott from the International Space Station," a faint voice responded. "Do you read me?"

Everyone's jaw dropped.

"Please repeat," the chief said.

"This is the International Space Station. Do you read me?"

"Yes, yes, we hear you," the chief stated as the communication became stronger. "What is your condition?"

"We're all fine now," Dorothy said. "But we believe we are now moving into an unusual orbit around the sun. It appears that we are moving into an orbit almost identical to that of Halley's Comet."

"Good grief," the chief said. "Don't jump to that conclusion. Let us look into it first. We are not sure if you are any safer there than we are here on Earth. Nevertheless, we are elated to know you are alive."

"You can thank Walter for our survival. He kept us busy figuring out ways to survive and to not give up."

"I would expect nothing less of him. What is the status of the station now?"

"It has suddenly started to operate properly. Until just a few moments ago, the autopilot system had kept all systems completely shut down, including all external communications. Ever since we initiated the earth orbit maneuver, the autopilot system cut off communication to stabilize the ISS, and we have not been able to override it."

"We now believe we know what happened with the AutoNav," the chief said.

"Here's what happened to us," Walter interjected. "Our instruments have been showing the earth moving in an unusually wide orbit around the sun. Now our orbit console shows it is back in its normal orbit. What the heck is happening back on Earth?"

"Are you serious? Did I hear you say that your readings show a normal earth orbit?" the captain shouted.

"Yes. The earth suddenly shows a normal orbit. No doubt about it. Can you please tell us what is going on?"

Everyone in the command center jumped out of his chair and screamed, "It worked! It worked!"

As word quickly spread, euphoria pervaded the room. The good news soon reached all available closed-circuit lines. Many world communication lines were just being restored, and people began sharing the news from the space station. People all over the world screamed, "It worked! It worked! It worked!" in hundreds of different languages.

However, one NASA communication coworker received some distressing news. The space station was indeed moving into an uncorrectable orbit around the sun. It appeared the ISS was indeed following a giant orbit similar to that of Halley's Comet. It would slingshot around the sun and far into outer space and return in roughly 122 years. If the ISS crewmembers were not killed by the solar slingshot effect, they would be catapulted back into outer space into an immense orbit. NASA had to break the bad news to the ISS members.

The Final Outcome

Dorothy began to sob. After a few moments, Walter spoke. "Please send this brief message in case this is our final communication," Walter said. "Tell Jason and Britney we love them very much. Tell them that their mother and I have achieved our lifelong dream of being on the space station together. Tell them that whatever their goals and ambitions, just strive to be the best. Tell them we will always love them. Tell them to follow their dreams."

"Wait a moment," the NASA captain said. "If the earth is back in its correct orbit, all you need to do is hit the reset button on the AutoNav and it will self-correct. So you can tell your kids those moving words in person. Consequently, I need you to please report for your debriefing back here on Earth in seven days."

Jason, his fiancée Jane, his best buddy Austin, and his fiancée Tyresha, Jason's his sister Britney, her roommate Manuela and the other UCLA team members huddled together in the research lab basement heard the entire dialogue over the closed-circuit communication line and began shouting and frolicking with glee. After several minutes, they collapsed to the floor from sheer exhaustion. Soon they noticed from the monitors that the sporadic lightning and tornadoes began to subside and the sky scrolls began to dissipate. They also hoped the Gaia Theory was correct and the Earth's ecosystems would begin to self correct.

Austin watched Jason and Jane tightly embracing and thought about the unbelievable struggles and tragedies they had endured and future challenges of restoring the earth's environment.

Jane stared back at Austin for a long time and then smiled and whispered something to Jason. She then turned to Austin. "Jason and I want you to be the godfather of our baby."

Austin's mouth flew open as he stared at Jane. With his eyes filled with tears, he looked down to regain his composure. He then stood up, opened his arms to Jane and Jason, and bowed.

"I would be honored," he said. "Especially if I can tell the baby that his dad really did save the world."

The Final Outcome